INTRODUCTION

Everyday Earth Science is filled with a lot of handy, high-interest activities students can complete to learn about some important Earth science topics. Some of the many topics covered in this book include minerals (how to identify them, their value, Mohs Hardness Scale, in foods, composition of); earthquakes (faults, types, locating past quakes); rocks; volcanoes; geologic time periods; meteorology (storms, weather instruments, temperature, humidity, clouds, wind); space (sun, moon, planets, stars); and many, many more.

Easy to use, many of these reproducible activities are self-contained and can be completed by students in a short amount of time at their desks. Some activities have been designed to help students reinforce their information-gathering skills and require quick access to such classroom-based references as almanacs, atlases, encyclopedias, newspapers, and magazines. Still a few others offer students the opportunity to study a topic in a more in-depth manner. Some of the activities even build on each other and can be dispersed in intervals over a period of several days.

Each activity has been created for students to do on an individual basis. Students will thoroughly enjoy learning about Earth science topics and concepts as they work puzzles, fill in charts, complete experiments, make observations, compare things using Venn diagrams, examine food labels, locate places on a map, construct maps, and so much more.

Regardless of whether you are trying to supplement your science curriculum, give students practice in applying scientific concepts, or encourage students to connect their own experiences to Earth sciences content, you will be pleased as they find Earth science topics interesting, challenging, and meaningful.

Name_____   Date _____

MINERAL IDENTIFICATIONS

Minerals are the most common solid materials found on Earth. Minerals may vary in the way they feel and look. Some identifying characteristics of minerals are listed below. Use your text and reference materials to complete the mineral chart.

Mineral	Hardness	Specific Gravity	Streak Color	Luster
siderite	3.5–4		white	
gypsum		2.32		vitreous
kaolinite			white	dull
halite	2.5		white	
fluorite		3–3.3		glassy
calcite	3	2.7		
barite		4.3–4.6	white	
pyrite	6–6.5		green-black	
galena		7.4–7.6		metallic
magnetite			black	
topaz			colorless	glassy

The hardness of a mineral is measured by its ability to be scratched. Some identification tests can be made using common items.

Hardness

0–2.5	Mineral can be scratched by one's fingernail.
3	Mineral can be scratched by a copper penny.
5.5	Mineral can be scratched with a knife but not a penny.
5.5–6.5	Mineral will scratch glass.
6.5	Mineral can be scratched slightly with a file.
above 6.5	Mineral cannot be scratched with a file.

1. Which of the minerals above can be scratched by one's fingernail?

_____

2. Which of the minerals above can be scratched by a copper penny?

_____

3. Which of the minerals above will scratch glass?

_____

4. Which of the minerals above cannot be scratched by a file?

_____

     FS-10615 Everyday Earth Science

everyday earth science

MISSING MINERALS

Minerals are not found only on Earth. Mercury, Mars, Venus, and also our moon contain minerals. Locate and circle the hidden minerals in the puzzle below.

BARITE	CORUNDUM	GRAPHITE	MICA	SERPENTINE
BAUXITE	DIAMOND	HALITE	QUARTZ	SILVER
CALCITE	FELDSPAR	HORNBLENDE	RUTILE	TALC
CINNABAR	GOLD	KAOLINITE	SALT	TOPAZ

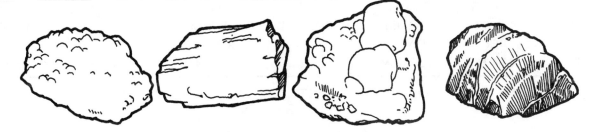

```
K E Z R M E R A P S D L E F K
T I T R U T I L E A L R D Z C
E N O I E T I N I L O A K O A
G O P I C H U Z V W G O R M R
G E A T L L C D A O N U I E W
R I Z H F F A A L P N C V N E
A Z S A A B P C T D U L L H T
P U H D L L F C U T I A B A I
H L O K D Z I M O S C T W L X
I H R C E N I T N E P R E S U
T J S Y N U O X E Q A N X O A
E D G A I I Y M Z E T I R A B
J T B Q L Z T R A U Q M J C C
N A M U F T R N L I G N I I E
R E H O R N B L E N D E S M B
```

2

Name_____ Date_____

Economically Important Minerals

Many minerals are mined for their economic importance in the manufacture of jewelry, electronic equipment, building materials, medicines, cosmetics, and pottery. These minerals are classified into different groups. Some of these groups are listed below the circles.

Research the minerals in the box to discover the class to which they belong. Write each mineral in the correct circle.

IRON	OLIVINE	MANGANITE	GRAPHITE	BORNITE	MALACHITE	SULFUR
BARITE	FLUORITE	GARNET	APATITE	PYRITE	FELDSPAR	HALITE
SODA NITER	BAUXITE	HEMATITE	DIAMOND	TURQUOISE	GYPSUM	BORAX
CALCITE	NITER	BERYL	GALENA	QUARTZ	MICA	GOLD
CUPRITE	COPPER	PLATINUM	RUTILE	WOLFRAMITE	DOLOMITE	SILVER

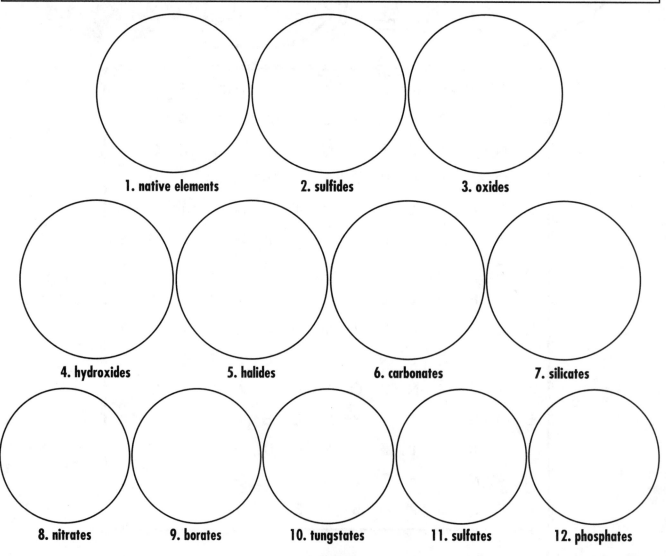

1. native elements   2. sulfides   3. oxides

4. hydroxides   5. halides   6. carbonates   7. silicates

8. nitrates   9. borates   10. tungstates   11. sulfates   12. phosphates

FS-10615 Everyday Earth Science

THE MOHS HARDNESS SCALE FOR MINERALS

In 1822, Friedrich Mohs, a German mineralogist, designed a hardness scale for minerals. Ten minerals are used for comparisons in the hardness of other minerals.

Unscramble the letters below to identify the 10 minerals in the scale. Print the words in the spaces.

1. L C T A   ___ ___ ___ ___

2. S P Y M U G   ___ ___ ___ ___ ___ ___

3. T I L E C A C   ___ ___ ___ ___ ___ ___ ___

4. T I R E L U F O   ___ ___ ___ ___ ___ ___ ___ ___

5. T A P I T E A   ___ ___ ___ ___ ___ ___ ___

6. D P A S F E R L   ___ ___ ___ ___ ___ ___ ___ ___

7. Z A R Q U T   ___ ___ ___ ___ ___ ___

8. P O Z A T   ___ ___ ___ ___ ___

9. D O C R U M U N   ___ ___ ___ ___ ___ ___ ___ ___

10. M I N D A D O   ___ ___ ___ ___ ___ ___ ___

After you have identified the 10 minerals, use reference materials to describe some of the physical properties of each mineral.

1. _____
2. _____
3. _____
4. _____
5. _____
6. _____
7. _____
8. _____
9. _____
10. _____

FS-10615 Everyday Earth Science

EVERYDAY EARTH SCIENCE

EVERYDAY EARTH SCIENCE

BEAUTIFUL GEMSTONES

Gemstones are all minerals and stones that are used for jewelry and other decorative purposes. Complete the grid with the names of gemstones listed below.

AMETHYST	DIAMOND	LAPIS LAZULI	TOPAZ
AQUAMARINE	EMERALD	OPAL	TOURMALINE
BERYL	GARNET	RUBY	TURQUOISE
CAT'S-EYE	JADE	SAPPHIRE	ZIRCON

FS-10615 *Everyday Earth Science*

HIDDEN DIAMONDS

A diamond is the hardest naturally-occurring substance. It is the most enduring of all gemstones. See how many times you can find the word DIAMOND in the puzzle below. The word is written forward, backward, up, down, and diagonally. Also try to find the hidden word that is used for the measurement of a diamond's weight.

```
D  I  A  M  O  N  D  I  A  M  O  N  D  C  T  W
R  I  Q  I  V  B  W  N  N  O  G  P  I  Q  K  M
T  A  C  R  H  U  E  S  O  T  A  R  A  C  T  B
J  M  Y  F  V  T  S  D  W  M  X  Z  M  A  V  L
U  O  B  R  D  C  E  I  H  D  A  F  O  I  G  Y
D  N  J  K  I  H  D  R  G  L  H  I  N  M  J  L
N  D  S  X  A  J  N  P  R  T  O  S  D  Q  X  M
O  U  W  G  M  L  Z  X  N  Y  B  I  A  Z  D  C
M  E  G  L  O  N  D  N  O  M  A  I  D  M  W  K
A  O  M  P  N  Q  V  I  F  M  R  S  D  I  C  D
I  T  U  F  D  V  Y  A  O  X  W  I  B  C  Z  N
D  I  A  M  O  N  D  N  D  F  A  G  E  H  I  O
O  J  M  K  P  L  D  N  O  M  A  I  D  N  O  M
Q  P  N  R  T  X  S  Q  O  Y  E  U  W  D  V  A
Z  A  C  B  F  H  E  N  J  D  G  O  R  K  Q  I
D  N  O  M  A  I  D  L  I  N  P  S  U  M  T  D
```

EVERYDAY EARTH SCIENCE

MINERALS IN FOODS

Minerals are chemical substances which appear naturally in the crust of the Earth and in foods. Use reference materials on nutrition and food labels to identify foods which contain the minerals listed in the chart below.

MINERAL	FOODS CONTAINING THE MINERAL
CALCIUM	
SODIUM	
POTASSIUM	
IRON	
PHOSPHORUS	
IODINE	
ZINC	
COPPER	
MAGNESIUM	

FS-10615 Everyday Earth Science

everyday earth science

METALS FROM MINERALS

Metal forms a large part of Earth. Earth's crust is made up of certain metals. Identify the metals below from the clues given.

1. I am obtained from the mineral bauxite.
   I am shiny and often seen as rolls of thin sheets.
   Many soft drink cans are made from me.
   I am used to make canoes, chairs, windows, and siding for houses.
   My chemical symbol is Al.
   What metal am I?

   _____

2. I am found in the minerals limonite, hematite, and magnetite.
   I am used in the manufacture of steel.
   I rust easily.
   I am also found in "fool's gold," or pyrite.
   My chemical symbol is Fe.
   What metal am I?

   _____

3. I am found in the mineral galena.
   I am used in pipes, batteries, and glass.
   I was once used in many paints.
   I am used to protect people from the effects of X-rays.
   My chemical symbol is Pb.
   What metal am I?

   _____

4. I am found in the mineral cinnabar.
   I remain liquid over a wide range of temperatures.
   I am shiny.
   I am used in many thermometers.
   My chemical symbol is Hg.
   What metal am I?

   _____

5. I am found in the mineral chalcopyrite and in others.
   I am used in most electrical wires.
   I am used in many coins.
   I am mixed with other metals to make bronze.
   My chemical symbol is Cu.
   What metal am I?

   _____

6. I am found in the mineral chromite.
   I am used in cars and bathroom fixtures.
   I am very shiny and one of the hardest substances.
   I am used to make stainless steel.
   My chemical symbol is Cr.
   What metal am I?

   _____

7. I am found in the mineral cassiterite.
   I am used in welding solder.
   I was once widely used as roofing material for houses and barns.
   I am mixed with copper to make bronze.
   My chemical symbol is Sn.
   What metal am I?

   _____

Name_____ Date _____

Tectonic Plates of the Earth

Earth's surface consists of about 20 rigid plates that move slowly past one another. Label the tectonic plates of Earth on the map below.

EURASIAN PLATE   INDIAN PLATE   PHILIPPINE PLATE
PACIFIC PLATE   NORTH AMERICAN PLATE   NAZCA PLATE
SOUTH AMERICAN PLATE   AFRICAN PLATE   ANTARCTIC PLATE

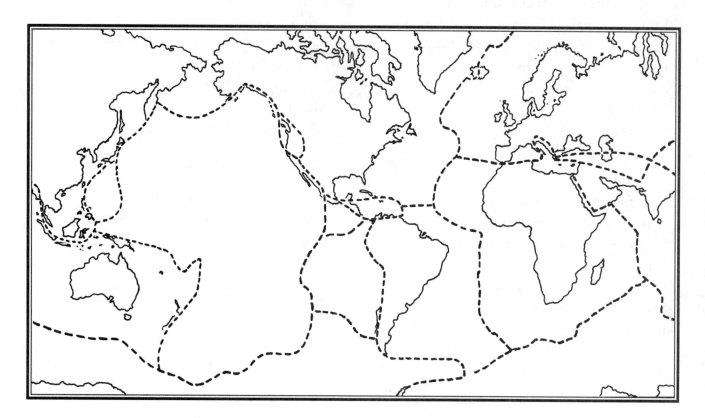

Write what you think (or know) an earthquake would do to your city/town.

_____

_____

_____

FS-10615 Everyday Earth Science

ALL ABOUT EARTHQUAKES

An earthquake is the sudden shaking of the ground that occurs when masses of rock change positions below Earth's surface. Learn more about earthquakes by reading the clues below. Locate the term in the magic square that matches each clue. Then write the number of the clue in the space. By recording all of the correct numbers, you will have produced a magic square. When you add the numbers across, down, or diagonally, you should get the same answer. The four squares in each corner of the big square and the four squares in the center of the big square will also give you the same answer when added together.

fault	San Francisco	strike-slip fault	focus
_____	_____	_____	_____
normal fault	Richter scale	primary waves	Buffalo, NY
_____	_____	_____	_____
secondary waves	surface waves	oil and fossils	epicenter
_____	_____	_____	_____
reverse fault	San Andreas Fault	seismograph	seismologist
_____	_____	_____	_____

1. a fracture within Earth where rock movement occurs

2. an instrument used to measure earthquakes

3. a large fault in California

4. the point in Earth where seismic waves originate

5. the point on Earth's surface directly above the focus

6. a numerical scale used to express the strength of an earthquake

7. seismic waves from the focus that are compressional

8. seismic waves from the focus that are perpendicular to this motion

9. location of the National Center for Earthquake Engineering Research

10. the most powerful shock waves from an earthquake

11. sometimes located in Earth by seismic waves from explosions

12. rock above a fault that moves downward

13. rock above a fault that moves upward

14. rocks that move in opposite horizontal directions

15. city which had major earthquakes in 1906 and 1989

16. scientist who studies earthquakes

EVERYDAY EARTH SCIENCE

THE SCIENCE OF EARTHQUAKES

An earthquake is a sudden shock of Earth's surface. Identify the name of the study of earthquakes by reading the clues below and writing the answers. The circled letters will spell out the name of this science. Print the name at the bottom of the page.

1. large ocean waves created by an earthquake
   __ Ⓞ __ __ __ __ __

2. These waves, created by the earthquake, are the strongest at the epicenter.
   __ Ⓞ __ __ __ __ __

3. the area on the surface of Earth directly above the occurrence of the earthquake
   __ __ Ⓞ __ __ __ __ __ __

4. famous earthquake fault in California
   Ⓞ __ __ __ __ __ __ __ __

5. the instrument used to record earthquake waves
   __ __ __ __ Ⓞ __ __ __ __ __ __

6. the origin of an earthquake under the surface of Earth
   __ __ Ⓞ __ __ __ __

7. a breaking point in layers of Earth
   __ __ __ __ Ⓞ

8. the vibrational tremors sent out from an earthquake
   __ __ __ Ⓞ __ __ __ __ __ __ __

9. the name given to the area around the Pacific Ocean in which many earthquakes occur
   __ __ __ __ Ⓞ __ __ __ __ __

10. the fastest waves from an earthquake; also called push waves
    __ __ __ __ __ __ Ⓞ

The science of the study of earthquakes is __ __ __ __ __ __ __ __ __ __ __ __.

THE WORLD'S MOST DESTRUCTIVE EARTHQUAKES

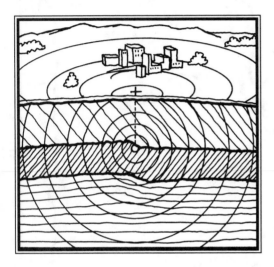

Earthquakes have been recorded on Earth for thousands of years. In addition to the damage to buildings, highways, and land, millions of people have been killed.

For this activity, you will need a copy of the World Map on page 13 and a globe or an atlas. Locate the countries in which the earthquakes below occurred. The estimated number of deaths from the earthquakes and the dates of the earthquakes are recorded. Place the number of the earthquake in its correct location on the map.

Earthquake	Date of occurrence	Number of estimated deaths
1. Syria	526	250,000
2. Iraq	847	50,000
3. Iran	856	200,000
4. India	893	180,000
5. Egypt	1201	1,000,000
6. Italy	1456	60,000
7. China	1556	830,000
8. Japan	1703	200,000
9. Ecuador-Peru	1797	41,000
10. Java	1883	100,000
11. Chile	1906	20,000
12. Turkey	1939	30,000
13. Nicaragua	1972	10,000
14. Guatemala	1976	23,000
15. Mexico	1985	4,200

Name _____ Date _____

Map of the World

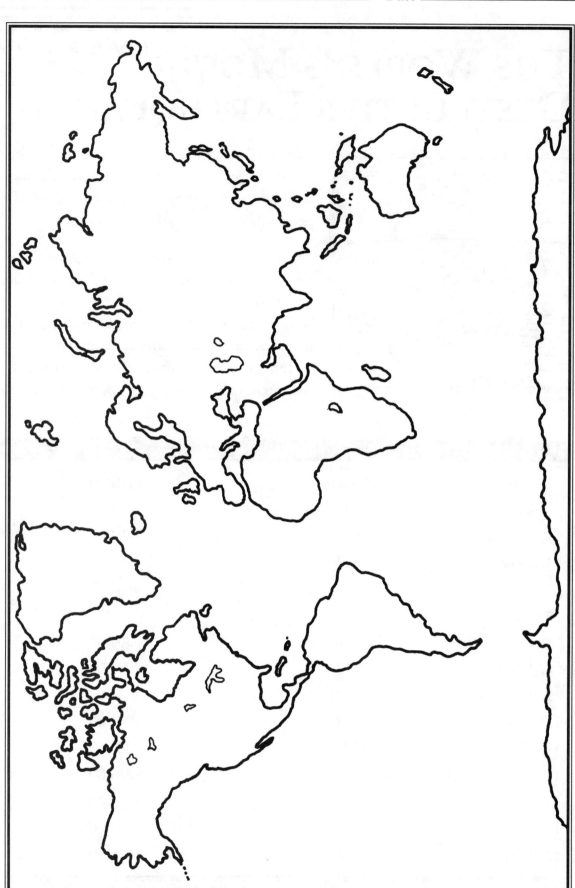

everyday earth science

Where Is the New Madrid Fault?

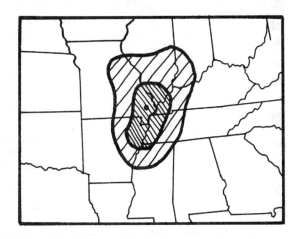

The largest earthquake in the United States occurred in the winter of 1811–1812 in an area along the New Madrid Fault. Use the map of the United States on page 15, a metric ruler, a thumbtack, a piece of corrugated cardboard, a 40-cm piece of string, and a pencil to locate the New Madrid Fault.

1. Place the map on a piece of corrugated cardboard.

2. Locate the following cities on the map: San Francisco, El Paso, Miami, Chicago, Atlanta, and Denver.

3. Calculate the actual distance in kilometers for each city below by using the scale, 1 centimeter = 210 kilometers. The distances between the New Madrid Fault and the cities listed are below.

4. Using the metric ruler, measure 13.0 centimeters east from San Francisco. Mark the distance using the pencil. Make a loop in the string, place the pencil in the loop, and hold the tip of the pencil on the mark you have made. Stretch the string to San Francisco and attach the string using the thumbtack. Draw a large arc on the paper using the pencil.

5. Repeat the measurements and sketches with the other five cities. Make sure the arcs you draw are large from top to bottom.

6. The location of the New Madrid Fault is the area where the six arcs seem to cross or intersect.

a. San Francisco	13.0 centimeters	_____ kilometers
b. El Paso	7.5 centimeters	_____ kilometers
c. Miami	7.0 centimeters	_____ kilometers
d. Chicago	3.0 centimeters	_____ kilometers
e. Atlanta	3.0 centimeters	_____ kilometers
f. Denver	6.0 centimeters	_____ kilometers

In what state is the New Madrid Fault located?

_____

Which cities in the U.S. would be greatly affected by an earthquake in this region today?

_____

_____

EVERYDAY EARTH SCIENCE

Map of the United States

CONTINENTS ON THE MOVE

A German geophysicist first proposed the idea that Earth's continents are slowly drifting apart. To find out who he was, use the letter grid and number codes below.

	1	2	3	4	5	6	7	8	9	10	11	12
k	E	N	D	P	W	A	C	R	L	F	M	R
j	X	O	S	C	L	E	M	B	R	U	X	Z
i	B	E	N	L	P	R	C	E	B	F	E	H
h	I	J	O	E	H	I	T	K	G	L	D	M
g	C	P	N	F	O	A	R	T	A	C	V	E
f	R	D	H	E	D	B	H	D	A	U	N	H
e	W	E	G	X	R	Y	S	Z	I	Y	J	R
d	R	F	Q	S	P	B	A	S	R	V	T	I
c	Y	C	G	X	O	E	D	L	E	K	G	W
b	K	I	J	Z	L	A	U	B	O	M	N	P
a	E	N	Z	N	Q	V	S	M	R	F	W	T

6g 4i 10k 12e 2e 11h   5k 6c 3c 8i 11b 1k 9d

3j 10f 3c 11c 1a 7e 8g 6j 11h   12a 3f 6k 8g

7d 5b 9k   2c 3h 11b 8g 1h 4a 4h 11b 11d 3j

5c 3i 2c 12g   12c 12g 5e 4h   9f

7a 10j 5d 1k 6i 1g 9b 4a 12a 9e 3g 9c 2a 8g

7i 6k 5b 9k 4f 8f   4k 7d 11b 3e 7d 12g 6k .

EVERYDAY EARTH SCIENCE

LAND FEATURES OF THE EARTH

Earth is covered with many fascinating features. Read each clue below to identify a land feature of Earth. Then find the word in each puzzle grid by tracing the letters with straight lines. No lines will cross. The first one has been done for you.

1. outer layer of Earth

CRUST

```
A   T   E   L   B
N   S   A   V   N
C   O   U   D   F
P   R   I   M   J
S   N   E   H   L
```

2. thick layer of solid rock below the crust

_____

```
P   O   L   B   M
E   U   R   A   C
L   S   N   V   N
D   T   S   E   F
R   A   C   F   H
```

3. center of Earth

_____

```
S   C   U   M   B
T   O   A   D   F
G   R   K   N   H
I   E   O   S   P
M   J   R   T   L
```

4. a hollow or natural passage under or into Earth, opening to the surface

_____

```
C   O   V   R   S
T   A   B   E   D
F   I   G   K   O
L   N   P   T   H
J   R   U   Y   M
```

5. flat land at the mouth of a river

_____

```
A   G   I   K   B
C   T   P   O   F
H   R   L   T   U
M   V   Y   E   J
S   N   A   G   D
```

6. low area between two mountains

_____

```
C   M   U   B   F
O   T   G   H   L
R   N   L   O   R
T   A   S   L   Y
V   U   P   W   E
```

7. large areas of Earth's crust that are pushed up

_____

```
T   B   S   C   M
R   N   W   O   D
I   L   U   E   P
A   N   O   R   S
T   U   W   A   C
```

8. a mountain that emits hot gases and lava

_____

```
D   A   M   B   E
N   F   C   G   K
O   I   O   L   R
J   V   P   T   H
L   E   S   A   N
```

9. a gorge cut through the land by water and wind

_____

```
C   N   B   L   M
O   U   D   A   E
T   Y   O   V   F
L   I   N   P   I
B   H   A   C   L
```

17

GLACIERS

Glaciers are thick masses of ice created by the accumulation and crystallization of snow. Match the clues about glaciers with the terms below.

1. _____ VALLEY GLACIER

2. _____ CIRQUE

3. _____ CONTINENTAL GLACIER

4. _____ CREVASSE

5. _____ DRUMLIN

6. _____ END MORAINE

7. _____ ESKER

8. _____ FIORD

9. _____ KETTLE

10. _____ PLUCKING

11. _____ ROCK FLOUR

12. _____ SURGE

13. _____ TARN

14. _____ TILL

A. material deposited directly by a glacier

B. glacier generally confined to mountain valleys

C. a crack in the glacier caused by movement

D. rapid movement of a glacier

E. the process whereby a glacier loosens and lifts rocks into the ice

F. pulverized rock caused by a glacier's abrasion

G. a bowl-shaped depression at the head of a glacial valley

H. a small lake formed after a glacier has melted away

I. a U-shaped depression formed by a glacier below sea level in a river valley that is flooded by the ocean

J. massive accumulations of ice that cover a large portion of a landmass

K. a hilly ridge of material formed at the end of a valley glacier

L. an oval-shaped hill consisting of rock debris

M. a depression left in part of a glacier formed by the melting of a block of ice

N. ridges of sand and gravel deposited by flowing rivers of melted ice through a glacier

EVERYDAY EARTH SCIENCE

MOLTEN ROCKS

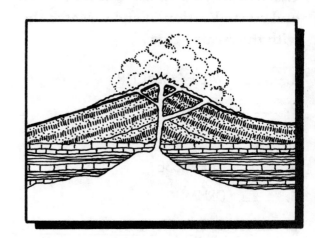

There are three main classes of rocks. Sedimentary and metamorphic are two classes. The other class of rocks in the Earth's crust formed from cooled lava or magma. The lava came to the surface of Earth, and the magma solidified.

To find out this last class of rocks, correctly fit the rocks listed below into the spaces. The circled letters will spell out the other class of rocks to which all of these rocks belong.

BASALT   DIORITE   FELDSPAR   GRANITE
OBSIDIAN   OLIVINE   QUARTZ

1. _____ _____ _____ _____ _____ _____ _____
2. _____ _____ _____ _____ _____ _____ _____ _____
3. _____ _____ _____ _____ _____ _____ _____
4. _____ _____ _____ _____ _____ _____ _____
5. _____ _____ _____ _____ _____ _____ _____
6. _____ _____ _____ _____ _____ _____ _____
7. _____ _____ _____ _____ _____ _____ _____
8. Class of rocks: _____ _____ _____ _____ _____

9. Use a reference book to identify some of the major uses of these rocks.

_____

_____

_____

10. Where are these rocks found in the world?

_____

_____

_____

everyday earth science

WHICH ROCKS ARE THESE?

The hard, solid part of the Earth is rock. Three main kinds of rocks are igneous, sedimentary, and metamorphic. To learn more about rocks, match the names of the rocks to their definitions.

1. _____ SANDSTONE

2. _____ SHALE

3. _____ CONGLOMERATE

4. _____ BRECCIA

5. _____ LIMESTONE

6. _____ SEDIMENTARY ROCKS

7. _____ BITUMINOUS COAL

8. _____ GEODES

9. _____ GNEISS

10. _____ SCHIST

11. _____ ANTHRACITE COAL

12. _____ MARBLE

13. _____ QUARTZITE

14. _____ SLATE

15. _____ METAMORPHIC ROCKS

A. a metamorphic rock which can be formed from limestone

B. a sedimentary rock of plant origin

C. the most common metamorphic rock made from granite

D. a metamorphic rock made from sandstone

E. a sedimentary rock made from cemented sand

F. rocks changed in form from sedimentary or igneous rocks

G. hard metamorphic rock made from soft coal

H. cemented mud, clay, or silt

I. coarse sedimentary rock made up of pebbles or boulders

J. metamorphic rock made from shale

K. sedimentary rock composed of rough and angular fragments of rock

L. metamorphic rock made from schist

M. sedimentary rock made from the shells and skeletons of plants and animals

N. hollow spheres in limestone

O. layered rocks made from sediments

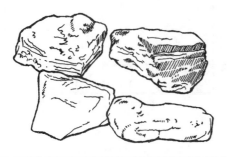

EVERYDAY *earth science*

Igneous Rocks

The three kinds of rocks are igneous, metamorphic, and sedimentary. Igneous rocks form from magma. Magma is melted rock that is located deep in Earth. Complete the grid by locating the names of the igneous rocks listed below.

ANDESITE

BASALT

DIORITE

DUNITE

FELSITE

GABBRO

GRANITE

OBSIDIAN

PEGMATITE

PERIDOTITE

PUMICE

QUARTZ

RHYOLITE

SYENITE

TUFF

Name_____   Date_____

SEDIMENTARY ROCKS

The three kinds of rocks are igneous, metamorphic, and sedimentary. Sedimentary rocks are made of materials that once were part of older rocks or of plants and animals. Complete the grid by locating the names of the sedimentary rocks listed below.

CHALK	COQUINA	LIMESTONE
CLAY	DOLOMITE	SANDSTONE
COAL	GRAYWACKE	SHALE
CONGLOMERATE	HALITE	SILTSTONE

FS-10615 Everyday Earth Science

METAMORPHIC ROCKS

The three kinds of rocks are metamorphic, sedimentary, and igneous. Metamorphic rocks are rocks that have changed appearance and sometimes have changed in their mineral composition. Complete the grid by locating the names of the metamorphic rocks listed below.

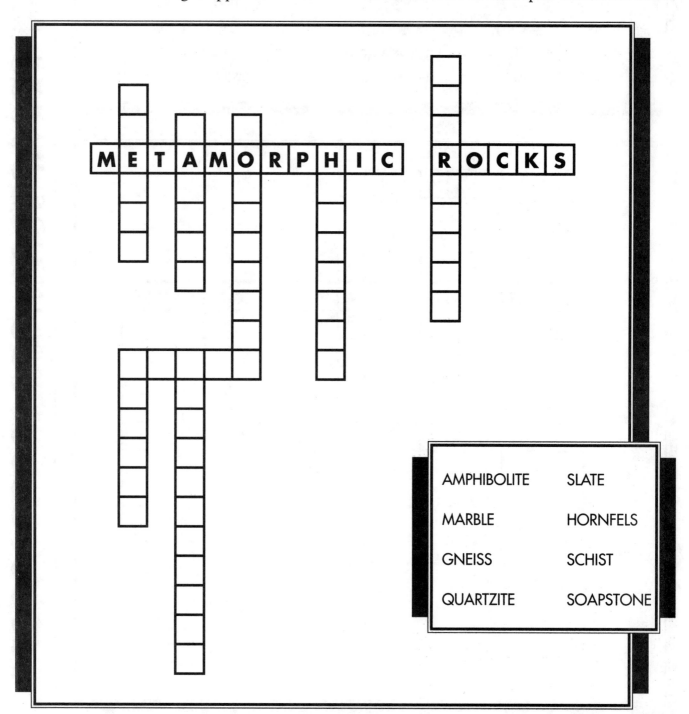

AMPHIBOLITE	SLATE
MARBLE	HORNFELS
GNEISS	SCHIST
QUARTZITE	SOAPSTONE

PUZZLE OF THE EARTH

See how much you know about rocks. Solve this crossword puzzle using the clues and the terms below.

BANDED	EXTRUSIVE	INTRUSIVE	METAL	PUMICE
CHALK	GRANITE	LAVA	METAMORPHIC	ROCK
COAL	HALITE	MAGMA	NONBANDED	SEDIMENTARY
CRUST	IGNEOUS	MARBLE	ORES	SLATE

ACROSS

2. rock formed from layers of rock pieces or sand cemented together

5. hot, liquid rock breaking through the surface of Earth

8. magma that cools inside Earth

11. rocks that contain useful metals

13. metamorphic rock, such as marble, that has no noticeable layers

14. rock made from other rocks by heat and pressure

17. a solid made up of different minerals

20. a metamorphic rock made from limestone

DOWN

1. metamorphic rock that has definite layers

3. igneous rocks that form on Earth's surface

4. a pure element such as iron, lead, gold, and silver

6. hot, liquid rock found beneath the crust of Earth

7. metamorphic rock made from shale

9. rocks formed from cooled magma

10. sedimentary rock made from compressed plants

12. an igneous rock used for buildings and monuments

15. sedimentary rock made of salt

16. the outermost layer of Earth

18. sedimentary rock made of the skeletons of marine animals

19. porous igneous rock used in abrasives and scouring soaps

EVERYDAY EARTH SCIENCE

GEOTHERMAL ENERGY FROM EARTH

Geothermal power is generated wherever water comes in contact with hot rocks below Earth's surface and turns into steam. Some geothermal power plants produce electricity more cheaply than ordinary power plants. Words related to geothermal power are hidden in the puzzle below. Locate and circle the words.

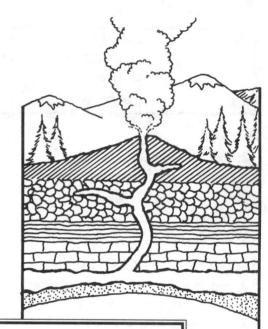

MUD POT	GEYSER	INNER CORE
OLD FAITHFUL	STEAM	OUTER CORE
MANTLE	ENERGY	UNDERGROUND
GEOTHERMAL	HEAT	ICELAND
PRESSURE	CRUST	YELLOWSTONE

```
O U T E R C O R E I N N O Q X
Y E O L O F A V P C O M S U R
B E P U V A G T O E R L E B O
J N L F G I M N R L A U L C L
C E R L T K P A U A S L S F A
E R M B O G R M W N T T I T M
R G U A C W E U A D E L B H R
O Y D G H C S W T E R L L D E
C O P R E D S T E A T F A E H
R G O E L K U C O F I S B E T
E L T N A M R E B N J J A Z O
N A K B W T E A K L E T R S E
N T A D N U O R G R E D N U G
I U R E S Y E G F L P S Z O U
Z O L D F A I T H F U L H R O
```

everyday earth science

FIVE TYPES OF MOUNTAINS

Geologists classify mountains into five basic types: dome, fold, fault-block, volcanic, and erosion. Make a sketch of each type below.

DOME MOUNTAINS
Black Hills of South Dakota
Weald Mountains in England

FOLD MOUNTAINS
Appalachian Mountains in the eastern U.S.
Alps in Europe, Himalayas of Asia

FAULT-BLOCK MOUNTAINS
Teton Range in Wyoming
Wasatch Range in Utah
Harz Mountains in Germany

VOLCANIC MOUNTAINS
Mount St. Helens in Washington
Mount Fuji in Japan

EROSION MOUNTAINS
Catskill Mountains in New York

FS-10615 Everyday Earth Science

Name_____ Date _____

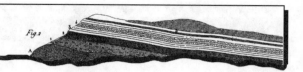

Volcanoes

A volcano is an opening in Earth's surface through which gases, lava, and ash erupt. To learn more about volcanoes, complete the crossword puzzle below.

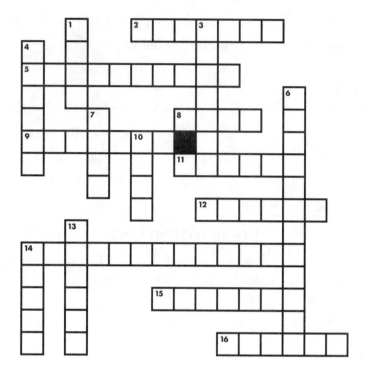

CALDERA   CONE   CRATER

DIKE   DORMANT   EXTINCT

GEYSER   LAVA   MAGMA

MOUNT ST. HELENS   PUMICE   PYROCLASTICS

RING OF FIRE   SHIELD   STRATO

TSUNAMI   VENT

ACROSS

2. a large, crater formed by the collapse of an overlying volcanic cone

5. the range around the Pacific Ocean where volcanoes mainly occur

8. fluid rock that pours from a volcano

9. a volcano that is not erupting and is not likely to erupt in the future

11. a composite volcano composed of alternating layers of lava and pyroclastic material

12. type of volcano that has a broad profile, such as Mauna Loa and Mauna Kea

14. the volcano that erupted in the state of Washington in 1980

15. a large seismic sea wave caused by a volcanic eruption or earthquake

16. groundwater which can be heated by volcanic activity and produces a hot-water fountain that spouts, such as Old Faithful

DOWN

1. an opening in the Earth's surface through which gases and lava may escape

3. an inactive volcano which is likely to erupt in the future

4. depression which can be caused by the collapse of a volcano

6. various-sized particles ejected by a volcano

7. a body of molten rock injected into a fissure in Earth

10. A cinder _____ volcano is one that has a conical shape and is composed mostly of cinder-sized pyroclastics.

13. a light, glassy rock formed from a frothy lava

14. molten rock inside Earth

FS-10615 Everyday Earth Science

VOLCANO ELEVATIONS

The elevations of some of the major volcanoes of the world are stated below. Cut out the strips, rank them in order from the lowest elevation to the highest elevation, and glue the strips to a sheet of construction paper. You can also identify the locations of these volcanoes on a globe or world atlas.

Mt. Kilimanjaro, Tanzania	5895 meters	Colima, Mexico	4100 meters
Mauna Kea, Hawaii	4205 meters	Mount Baker, WA	3285 meters
Slamet, Java	3428 meters	Paricutín, Mexico	2808 meters
Mauna Loa, Hawaii	4169 meters	Pacaya, Guatemala	2552 meters
Mount Fuji, Japan	3776 meters	Katmai, Alaska	2046 meters
Marapi, Sumatra	2891 meters	Pelee, Martinique	1397 meters
Mayon, Philippines	2461 meters	Kiska, Alaska	1220 meters
Kilauea, Hawaii	1222 meters	Guallatiri, Chile	6058 meters
Krakatau, Indonesia	813 meters	El Misti, Peru	5825 meters
Mount Etna, Italy	3390 meters	Tupungatito, Chile	5638 meters
Poas, Costa Rica	2708 meters	Ruiz, Colombia	5400 meters
El Chichón, Mexico	1060 meters	Sangay, Ecuador	5229 meters
Arenal, Costa Rica	1633 meters	Tolima, Colombia	5214 meters
Surtsey, Iceland	173 meters	Guagua Pichincha, Ecuador	4793 meters
Beerenberg, Norway	2277 meters	Galeras, Colombia	4276 meters
Askja, Iceland	1570 meters	Mount Shasta, California	4317 meters
Vesuvius, Italy	1277 meters	Vulcano, Italy	195 meters
Thera, Greece	564 meters	Stromboli, Italy	924 meters
Mount Rainier, WA	4392 meters	Fuego, Guatemala	3763 meters

FS-10615 *Everyday Earth Science*

Name_____ Date _____

GEOLOGICAL FORMATIONS

Choose one of the geological formations below to research. Write 12 clues/facts about it in the space provided. Read them to the class. See if the class can guess what it is.

AYERS ROCK, AUSTRALIA

BRYCE CANYON, UTAH

NEW MADRID FAULT

CARLSBAD CAVERNS, NEW MEXICO

NIAGARA FALLS

OLD FAITHFUL GEYSER

PIKES PEAK

GLACIER NATIONAL PARK, MONTANA

ROYAL GORGE

ICELAND

SAN ANDREAS FAULT

KILAUEA, HAWAII

SURTSEY ISLAND

KRAKATAU

YELLOWSTONE NATIONAL PARK

LURAY CAVERNS, VIRGINIA

ZION CANYON

MAMMOTH CAVE, KENTUCKY

MATTERHORN, SWITZERLAND

MAUNA KEA, HAWAII

MAUNA LOA, HAWAII

MT. ETNA, SICILY

MT. EVEREST

MT. FUJI, JAPAN

MT. KILIMANJARO, TANZANIA

PARICUTÍN

MT. ST. HELENS, WASHINGTON

CLUES/FACTS

1. _____

2. _____

3. _____

4. _____

5. _____

6. _____

7. _____

8. _____

9. _____

10. _____

11. _____

12. _____

Name_____   Date_____

THE GEOLOGICAL TIME SCALE

Identify the name of the oldest era on the geological time scale by solving the clues below. The circled letters will spell out the name of this era. Print the name at the bottom of the page.

1. the measurement of time in which rocks are deposited

　◯　__ __ __ __ __ __ __

2. microscopic organisms which lived in this era

　__ __ __ __ __ ◯ __ __

3. the measurement of time for each geological era

　__ ◯ __ __ __

4. The bottom of this gorge in Arizona was formed during this era.

　__ __ __ __ __ ◯ __ __ __ __

5. These land features were pushed up on the crust of Earth.

　__ __ __ __ __ ◯ __ __

6. The end of this era was about 600 _____ years ago.

　◯ __ __ __ __ __ __

7. This era began about 4.5 _____ years ago.

　◯ __ __ __ __ __ __

8. These objects from space have hit Earth and are about the same age as Earth.

　__ __ __ __ __ ◯ __ __ __ __

9. These preserved remains of plants and animals formed after this era.

　__ __ __ __ __ ◯ __ __

10. Write the name of the era that followed this one.

　__ ◯ __ __ __ __ __

11. This era lasted the _____ of all the eras of geologic time.

　__ __ __ ◯ __ __

Name of this era:　__ __ __ __ __ __ __ __ __

Name _____   Date _____

GEOLOGIC TIME

Create a geological time chart. This chart will help you understand the development of Earth and life on Earth. To create the chart, use geology reference materials to help you place the geologic time periods listed below in the correct order. Cut out the strips. Arrange them in order by placing the oldest period at the bottom and the most recent at the top. Glue them to a sheet of construction paper. Use the names of the four eras to separate the periods into four groups. For more fun, list what appeared on Earth during each period.

PERIODS		ERAS
SILURIAN	OLIGOCENE	CENOZOIC
PERMIAN	PLIOCENE	MESOZOIC
MISSISSIPPIAN	PALEOCENE	PALEOZOIC
PENNSYLVANIAN	MIOCENE	PRECAMBRIAN
CAMBRIAN	TRIASSIC	
HOLOCENE	ORDOVICIAN	
EOCENE	CRETACEOUS	
PLEISTOCENE	JURASSIC	
DEVONIAN		

everyday earth science

METEOROLOGY

Meteorology is the study of weather. Complete the grid below with the words related to meteorology.

ANEMOMETER

BAROMETER

CLOUD

CYCLONE

DEW

FRONT

HAIL

HURRICANE

MONSOON

RADAR

RAIN

SATELLITE

SLEET

SNOW

THERMOMETER

TORNADO

TYPHOON

WINDS

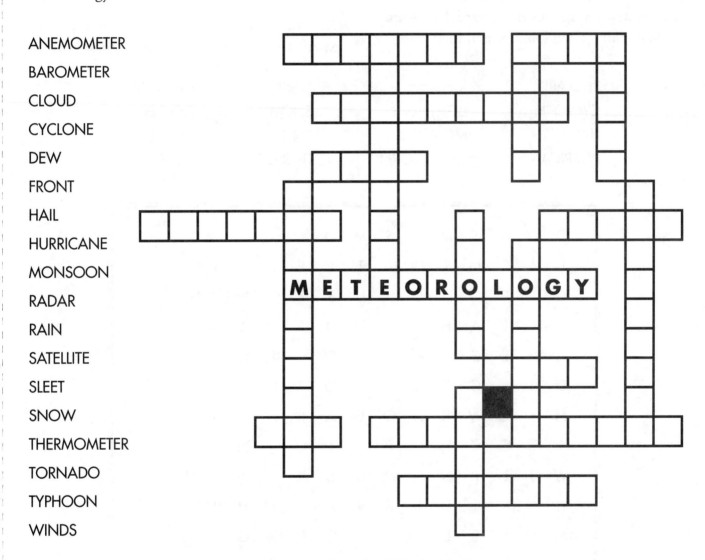

Define three words listed above with which you are not familiar.

1. _____

2. _____

3. _____

EVERYDAY *earth science*

Stormy Weather

Below is a list of stormy weather words. Locate and circle these words in the grid. The words may be written up, down, forward, backward, or diagonally. Then look for a hidden word that is a type of storm. Draw a box around this word.

On another sheet of paper, write about an experience you have had in one of these storms.

BLIZZARD	ICE	SLEET	TORNADO
CYCLONE	LIGHTNING	SNOW	TWISTER
HAIL	MONSOON	SQUALL	WATERSPOUT
HURRICANE	SANDSTORM	THUNDER	WIND

```
D  N  A  S  M  Q  S  M  O  L  T  E  A  R  E
B  A  C  M  R  O  A  T  L  E  S  L  E  E  T
C  L  E  G  I  D  N  H  B  F  O  N  J  L  O
V  L  I  A  H  P  D  U  D  S  U  W  Y  Q  R
A  H  E  Z  T  L  S  N  O  C  F  G  I  H  N
D  J  L  G  Z  L  T  D  Q  S  U  N  P  N  A
N  Y  Z  W  K  A  O  E  A  R  W  I  X  V  D
I  E  G  C  H  U  R  R  I  C  A  N  E  B  O
W  J  L  P  N  Q  M  D  O  K  T  T  Q  E  R
H  N  R  I  U  S  X  V  W  S  E  H  W  N  T
M  O  Y  E  D  A  E  O  G  D  R  G  C  O  B
P  O  Z  F  T  H  N  I  K  M  S  I  N  L  J
R  S  V  Q  O  S  H  E  A  R  P  L  T  C  U
Z  N  X  A  D  W  I  Z  C  G  O  B  F  Y  E
J  O  L  I  N  P  R  W  K  M  U  H  E  C  I
Y  M  T  S  U  D  O  Z  T  A  T  Y  W  U  S
```

Name_____ Date_____

A Magic Square of Weather

Below are words relating to weather. Write the number of the word which fits a clue in a box on the grid. If you have matched the correct numbers in all 16 squares, the sums of the rows, columns, and diagonals will be the same. This is called a magic square.

1. atmosphere
2. troposphere
3. ionosphere
4. ozone

5. jet streams
6. stratosphere
7. mesosphere
8. exosphere

9. wind
10. greenhouse effect
11. convection
12. sea breeze

13. land breeze
14. doldrums
15. trade winds
16. front

mass of air that surrounds Earth ____	air that rushes in from the north and south to warm the air along the equator ____	calm areas of Earth where there is little wind ____	a gas in the upper part of Earth's atmosphere ____
cold air from the ocean that moves into the warmer land ____	the zone of the atmosphere above the troposphere ____	the zone of the atmosphere above the stratosphere ____	a movement of air close to Earth's surface ____
the outer zone of Earth's atmosphere ____	air above Earth that is warmed by the reflection of the sun's rays and is prevented from easily passing back into space ____	transfer of heat by currents of air or water ____	strong, steady winds high in the atmosphere; used by pilots ____
cold air from land that moves out to warmer air over oceans ____	zone of the atmosphere which affects the transmission of radio waves ____	the zone of the atmosphere which is closest to the surface of Earth ____	the line along which air masses meet ____

What is the magic number for this puzzle? _____

Can you discover other number combinations in the puzzle which give you the same answer?

_____

_____

FS-10615 *Everyday Earth Science*

Name _____ Date _____

WEATHER-FORECASTING INSTRUMENTS

Observation stations are places where weather conditions on land are recorded. Many instruments are used to help people record weather information. Learn about some of them by identifying the correct weather instruments. Follow the directions after each answer and mark out letters in the grid. The remaining letters can then be written in the spaces to give you a hidden message about a certain weather instrument.

1. Wind speed is measured with a(n)
a. wind sock. Mark out all A's.
b. anemometer. Mark out all F's.
c. barometer. Mark out all G's.
d. psychrometer. Mark out all M's.

2. Wind direction is measured with a
a. weather vane. Mark out all K's.
b. radar. Mark out all C's.
c. barometer. Mark out all D's.
d. hygrometer. Mark out all H's.

3. Precipitation is measured with a(n)
a. anemometer. Mark out all P's.
b. pyschrometer. Mark out all R's.
c. barometer. Mark out all S's.
d. rain gauge. Mark out all X's.

4. Air pressure is measured with a
a. radiosonde. Mark out all B's.
b. rain gauge. Mark out all E's.
c. thermometer. Mark out all N's.
d. barometer. Mark out all J's.

5. Air moisture is measured with a(n)
a. anemometer. Mark out all I's.
b. radiosonde. Mark out all L's.
c. hygrometer. Mark out all Q's.
d. barometer. Mark out all O's.

6. Temperature is measured with a(n)
a. anemometer. Mark out all T's.
b. barometer. Mark out all U's.
c. thermometer. Mark out all Y's.
d. psychrometer. Mark out all V's.

F	K	W	E	J	Q	A	X	Y	K	T	F	H	Q	J	E	X	Q	R	Y	X	J	Y	Q	K
S	Y	F	A	Q	K	J	T	X	Y	E	L	J	L	F	I	K	T	Q	E	Y	K	S	J	F
J	A	X	R	K	E	F	Y	V	J	A	Q	L	Y	U	X	A	F	B	K	L	Y	J	E	K
F	Q	I	J	N	X	K	P	Y	Q	J	R	E	K	Y	D	I	C	X	T	I	J	K	N	G
K	H	Y	U	X	J	R	Y	K	R	X	I	Q	C	Y	J	A	F	N	K	Y	E	X	J	K
J	X	M	O	F	K	V	X	Y	J	E	Y	K	M	Y	X	E	Y	K	N	F	T	J	Q	S

Message:

__ __ __ __ __ __ __   __ __ __ __ __ __ __ __ __

__ __ __ __ __   __ __ __ __ __ __ __ __ __

__ __ __ __ __ __   __ __ __ __ __ __ __ __

__ __ __ __ __ __ __ __ __ __ __ __ .

Name_____ Date _____

WEATHER INSTRUMENTS

Weather conditions are measured using standard instruments. Find out what some of these instruments are by using the clues below to unscramble the letters of each weather instrument. The circled letters will then spell the source of all weather conditions on Earth.

1. an instrument carried aloft by a weather balloon to measure upper-level pressure, temperature, humidity, and winds

   O R I S E D N O D A   __ __ __ __ Ⓞ __ __ __ __

2. a type of radar that continuously measures the wind, moisture, and temperature of the upper atmosphere

   P L E D R O P   __ Ⓞ __ __ __ __

3. measures the ceiling or base height of cloud layers

   M O L I C E R E E T   __ __ __ Ⓞ __ __ __ __ __

4. measures precipitation in inches

   N A I R   A G G E U   __ Ⓞ __ __ __ __ __

5. measures the intensity of rainfall or snowfall

   A R D A R   __ __ __ __ Ⓞ

6. measures surface wind speeds

   E M O N E R T A M E   __ __ __ __ __ Ⓞ __ __ __

7. measures wind direction

   E V A N   __ __ Ⓞ __

8. measures air pressure

   R O B E T E R A M   __ __ __ __ Ⓞ __ __ __

9. measures temperature

   T R O M E M T H E E R   __ __ __ Ⓞ __ __ __ __ __ __

10. a special thermometer that measures temperature continuously

    G E R M A P T H O R H   __ __ __ __ __ Ⓞ __ __ __

11. measures relative humidity, vapor pressure, and dew point

    G R O T R Y E M E H   __ Ⓞ __ __ __ __ __ __ __

Answer: __ __ __ __   __ __ __ __ __

Name _____ Date _____

TEMPERATURE VARIATIONS

Temperature variations are caused by a variety of things. In this activity, you will measure temperature variations caused by different materials.

Materials Needed: 5 plastic cups with lids, 5 student thermometers, aluminum foil, white paper, black paper, cotton cloth, woolen cloth

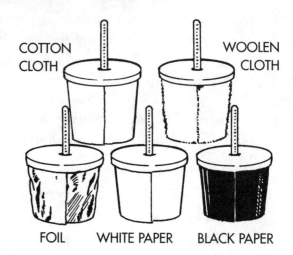

COTTON CLOTH   WOOLEN CLOTH

FOIL   WHITE PAPER   BLACK PAPER

1. Punch a small hole in each lid to allow a thermometer to be inserted.

2. Place a lid on each cup and insert the thermometer until the room temperature can be read slightly above the lid.

3. Wrap each cup in one of the different materials listed above, allowing the thermometer to be read.

4. Place all 5 cups in a sunny spot in your home or classroom. Make sure all 5 cups receive the same amount of sunlight.

5. Record the temperatures of the cups over a one-hour period.

Cup	Temperature at						
	Start	10 min.	20 min.	30 min.	40 min.	50 min.	60 min.
foil							
white paper							
black paper							
cotton cloth							
woolen cloth							

1. Which cup proved to have the least change in temperature? _____

2. Which cup proved to have the greatest change in temperature? _____

3. How would these results help you in choosing what to wear on extremely hot days? _____

RELATIVE HUMIDITY

Humidity is a measurement of the amount of water vapor in the atmosphere. Relative humidity is a comparison of this amount of vapor at a specific temperature to the total amount of water vapor the air could contain at that same temperature. For example, a relative humidity of 50% means that the air has only half as much water vapor as it could contain if it were completely saturated.

Relative humidity is often determined by two readings on a hygrometer, a dry-bulb thermometer reading, and a wet-bulb reading. The difference in the two readings is used along with the relative humidity chart depicted below. For example, if the dry-bulb reading is 58 and the wet-bulb reading is 48, the difference is 10 degrees. From the chart, the relative humidity would be 46%.

Use the relative humidity chart to find the percentages.

RELATIVE HUMIDITY (FAHRENHEIT)										
	Depression of Wet-Bulb Thermometer									
Dry Bulb	2	4	6	8	10	12	14	16	18	20
40	83	68	52	37	22	11	0			
42	85	69	55	40	26	16	0			
44	85	71	56	43	30	20	4	0		
46	85	72	58	45	32	23	8	0		
48	86	73	60	47	35	26	12	1	0	
50	87	74	61	49	38	29	16	5	0	
52	87	75	63	51	40	32	19	9	0	
54	88	76	64	53	42	35	22	12	3	0
56	88	76	65	55	44	37	25	16	7	0
58	88	77	66	56	46	39	27	18	10	1
60	89	78	68	58	49	41	30	21	13	5
62	89	79	69	59	50	43	32	24	16	8
64	90	79	70	60	51	45	34	26	18	11
66	90	80	71	61	53	46	36	29	21	14
68	90	80	71	62	54	48	38	31	23	16

DRY BULB	WET BULB	RELATIVE HUMIDITY
40	38	
48	34	
56	50	
60	46	
62	56	
66	64	
68	52	

Name _____   Date _____

A Station Model for Weather

The National Weather Service uses station models to represent the weather conditions in an area. Symbols and numbers are used to represent many aspects of weather including wind direction, precipitation, dew point, barometric pressure, and temperature.

Examine the station model below. The information shown includes the following:

temperature	45 degrees F
dew point	33 degrees F
wind direction	from NE
barometric pressure	259 (1025.9 mb)
precipitation in past 6 hours	.36 inches

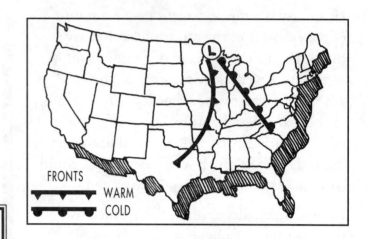

FRONTS

WARM
COLD

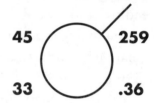

Use the data below to complete the station models.

MODEL	TEMPERATURE	DEW POINT	WIND DIR.	BAR. PRESSURE	PRECIPITATION
A	78	69	NW	241	.11
B	29	28	N	189	.87
C	98	67	SE	137	3.05
D	89	76	SW	238	1.56

A.

B.

C.

D.

FS-10615 Everyday Earth Science

Name_____   Date_____

ATMOSPHERIC CIRCULATION

There are five zones of atmospheric circulation on Earth. To find out what they are, begin with the letter D on the spiral and skip every other letter to spell out these five zones. Write the names of the zones in the spaces at the bottom of the page.

The five zones are: __ __ __ __ __ __ __ __ __

__ __ __ __ __ __   __ __ __ __ __

__ __ __ __ __ __ __ __ __ __ __ __ __ __ __ __ __

__ __ __ __ __ __ __

FS-10615 Everyday Earth Science

EVERYDAY EARTH SCIENCE

NAMES OF CLOUDS

A cloud is a mass of tiny water droplets or ice crystals that floats in the air. Identify the names of six clouds in the coordinate puzzle below. The first number of each number set represents the horizontal coordinate, and the second number of the set represents the vertical coordinate. For example, the number set 7-5 would be the letter T.

```
9  F  D  L  M  N  O  P  T  U
8  I  T  A  C  S  O  A  V  T
7  O  A  U  R  U  B  S  U  A
6  R  Y  M  A  L  I  D  U  T
5  E  L  G  T  J  F  T  C  L
4  S  K  U  S  M  U  I  H  A
3  U  N  Q  R  P  S  A  S  T
2  R  U  M  T  U  R  D  L  B
1  C  V  U  E  C  T  N  F  S
   1  2  3  4  5  6  7  8  9
```

1. ____ ____ ____ ____ ____ ____

   1-1   7-4   6-2   4-7   9-9   9-1

2. ____ ____ ____ ____ ____ ____ ____ ____ ____ ____ ____

   8-5   1-8   4-3   1-6   6-9   7-7   8-9   1-2   7-8   4-2   6-4   1-4

3. ____ ____ ____ ____ ____ ____ ____ ____ ____ ____ ____

   2-7   3-9   9-8   6-8   5-1   8-6   3-6   5-7   8-2   3-7   8-3

4. ____ ____ ____ ____ ____ ____ ____

   5-8   4-5   4-3   7-3   2-8   2-2   6-3

5. ____ ____ ____ ____ ____ ____ ____

   1-1   3-1   5-4   1-3   9-5   5-2   4-4

6. ____ ____ ____ ____ ____ ____ ____ ____ ____ ____ ____ ____

   5-9   6-6   3-2   9-2   6-9   4-4   9-6   4-3   9-7   7-5   9-9   6-3

FS-10615 Everyday Earth Science

EVERYDAY

earth science

WHAT DO YOU KNOW ABOUT TORNADOES?

Tornadoes are the most violent of all storms. Test your knowledge about tornadoes. Place a **T** before each true statement and an **F** before each false statement.

1. _____ Tornadoes occur only in the United States.

2. _____ Tornadoes are also called cyclones.

3. _____ In a tornado, the air spirals mostly vertically.

4. _____ The funnel cloud can be seen when it contains dust or debris.

5. _____ Tornadoes are very predictable.

6. _____ The central plains states experience the most tornadoes.

7. _____ If you are in a car during an approaching tornado, you should always try to outrun it.

8. _____ If you are in an open field during an approaching tornado, you should always seek a low-lying area and lie flat.

9. _____ You should take time to open all windows and doors during a tornado.

10. _____ Flying glass is a great danger in the home.

11. _____ Heavy rain and lightning often occur before and during a tornado.

12. _____ Most tornadoes last only a few seconds or minutes.

13. _____ A basement, utility room, or inside hallway are good places to seek shelter during a tornado.

14. _____ Intense cold fronts and squall lines create tornado conditions.

15. _____ A late afternoon calmness and a yellow sky are good warning signs that a tornado could occur.

16. _____ Tornadoes can skip or bounce from one site to another.

17. _____ Tornadoes can produce roaring sounds.

18. _____ Tornadoes over water are called waterspouts.

19. _____ The dark base cloud of tornado conditions is called the wall cloud.

20. _____ Most tornadoes hit small outlying cities rather than big cities.

21. _____ Fallen electric lines are an extreme danger after a tornado.

22. _____ A car is a very safe place during a tornado.

23. _____ The funnel of a tornado always touches the ground.

24. _____ Tornadoes always occur during the spring and summer.

EVERYDAY EARTH SCIENCE

A WINDY DAY

Research one of the special types of wind in the chart below.
Report your findings to the class and complete the chart.

NAME OF WIND	DEFINITION, SPECIAL FEATURES, OCCURRENCES
TRADE WIND	
GALE	
BORA	
KATABATIC	
FOEHN	
CHINOOK	
SANTA ANA	
MISTRAL	
SIROCCO	
ZEPHYR	
NORTHER	

everyday earth science

COMPARING TWO STORM SYSTEMS

Use the Venn diagram to compare the similarities and differences between tornadoes and hurricanes.

TORNADOES   HURRICANES

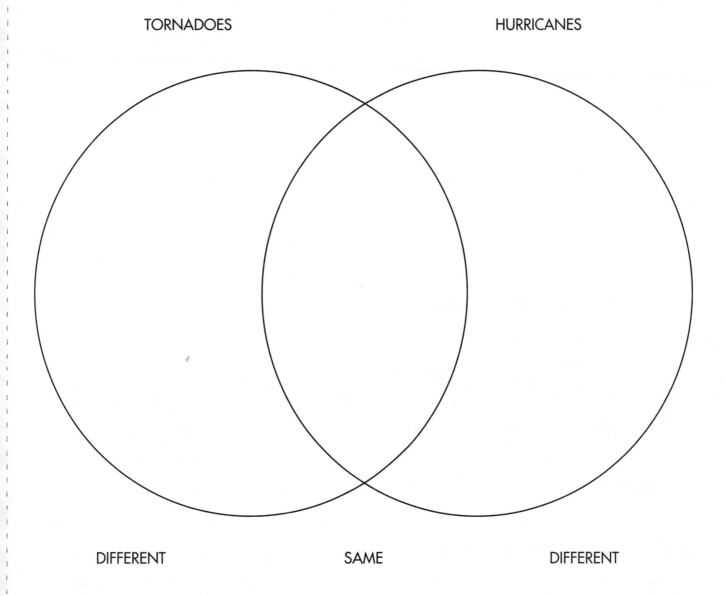

DIFFERENT   SAME   DIFFERENT

Name _____ Date _____

A Daily Weather Map

For this activity, you will need a daily weather map from a newspaper.
Cut out and glue the map in the space below. Then use the map to answer the questions.

1. What types of fronts are shown on the map? _____

_____

2. Where are the fronts located? _____

_____

3. In which states is precipitation shown? _____

_____

4. What types of precipitation are shown? _____

_____

5. Which states have the highest temperatures? _____

_____

6. Which states have the lowest temperatures? _____

_____

7. Which pressure systems are shown? _____ High _____ Low

A Concept Map of Meteorology

Examine some terms related to meteorology or weather below. First, identify the six major subheadings and write those terms in the circles. Then write the other terms as branches of these six subheadings.

RADAR
WEATHER MAP
STORM SYSTEMS
CIRRUS
CUMULONIMBUS
TORNADO
MONSOON
RAIN
IONOSPHERE
TROPOSPHERE

COLD FRONT
BAROMETER
HAIL
WIND VANE
SNOW
STRATOSPHERE
ISOBAR
ANEMOMETER
DEW
THERMOMETER

THUNDERSTORM
TYPES OF CLOUDS
MESOSPHERE
SLEET
WEATHER
    INSTRUMENTS
HYGROMETER
STRATUS
TYPHOON

TYPES OF
    PRECIPITATION
WARM FRONT
HIGH PRESSURE
LOW PRESSURE
HURRICANE
CUMULUS
LAYERS OF THE
    ATMOSPHERE
THERMOSPHERE

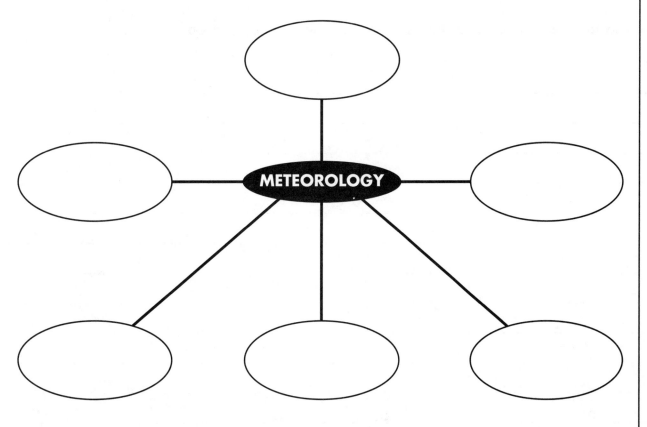

EVERYDAY EARTH SCIENCE

WEATHER WORDS IN SONG AND STORY

Think of titles of songs, books, stories, and movies that contain the following words:

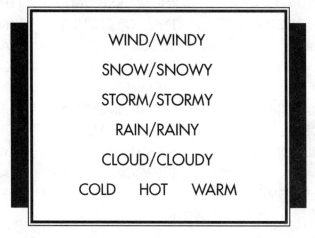

WIND/WINDY

SNOW/SNOWY

STORM/STORMY

RAIN/RAINY

CLOUD/CLOUDY

COLD   HOT   WARM

Write the titles below and underline the weather word(s) in each.

_____

_____

_____

_____

_____

_____

_____

_____

_____

_____

_____

FS-10615 Everyday Earth Science

E V E R Y D A Y

earth science

WEATHER TRIVIA

There is a lot to know about weather. See how much you know.
Place a **T** before each true statement below and on page 49 about weather.
Place an **F** before each false statement about weather.

1. _____ Fog is a cloud at ground level.

2. _____ A barometer is used to measure wind speed.

3. _____ Storms in the Pacific Ocean are called hurricanes.

4. _____ Hail is frozen rain.

5. _____ A barometer reading usually rises sharply before a storm.

6. _____ The greatest amount of gas in the air is oxygen.

7. _____ The troposphere is the area of the atmosphere closest to Earth.

8. _____ Hurricanes form over land.

9. _____ Hurricanes are named for men and women.

10. _____ Warm air can hold more water vapor than cold air.

11. _____ Sleet can be a mixture of rain and ice pellets.

12. _____ Acid rain forms when certain gases in the air mix with rain.

13. _____ Much of North America was covered with ice during the Ice Age.

14. _____ Dew is water vapor dropping from thunderstorms.

15. _____ Thunder is heard before the flash of lightning is seen.

16. _____ Lightning never hits tall trees.

17. _____ All life on Earth exists in the ionosphere.

18. _____ The amount of water in the air is called condensation.

19. _____ Evaporation occurs when liquid water changes to water vapor.

20. _____ A scientist who specializes in weather is called an archaeologist.

21. _____ Jet streams are high-altitude belts of high-speed winds.

22. _____ Winds blow clockwise in high pressure areas north of the equator.

23. _____ Condensation is the changing of ice into snow.

24. _____ Tornadoes usually form over land.

WEATHER TRIVIA Continued

25. _____ Weather conditions are caused by the sun, Earth, water, and air.

26. _____ Weather usually moves from east to west in the United States.

27. _____ A wind vane measures the speed of the wind.

28. _____ A rapidly falling barometer indicates the approach of a storm.

29. _____ In an open field during lightning, always seek shelter under a tree.

30. _____ The hygrometer measures the amount of dust in the air.

31. _____ The highest layer of the atmosphere is called the exosphere.

32. _____ Water heats up and cools down faster than land.

33. _____ The boundary between air masses is called a front.

34. _____ The layer directly above the troposphere is the ionosphere.

35. _____ Weather balloons carry instruments high into the atmosphere.

36. _____ The word RADAR stands for **RA**dio **D**etection **A**nd **R**ange.

37. _____ A rain gauge measures the amount of rainfall.

38. _____ The National Meteorological Service is located in Miami, Florida.

39. _____ NOAA stands for National Oceanic and Atmospheric Administration.

40. _____ Weather predictions, such as "red sky at night," are called folklore.

41. _____ Computers are extremely helpful today in weather forecasting.

42. _____ GOES satellites send weather pictures that are shown on television.

43. _____ Radar signals will pick up precipitation but not clouds.

44. _____ A psychrometer measures the amount of snowfall in 24 hours.

HELLO OUT THERE!

Our Earth and sun belong to a vast number of stars called the Milky Way Galaxy. The word *galaxy* comes from a Greek word meaning *milk*.

Use the code below to design a message you would send to outer space to tell any possible life forms in the Milky Way about Earth.

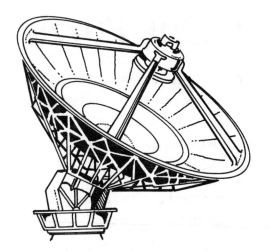

The code chart (A–Z).

1. Decode the message.

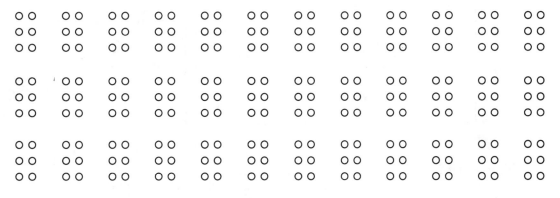

_____ _____ _____ _____ _____ _____ _____ _____ _____ _____ _____ _____

2. Design your own message to send. Exchange it with a classmate to decode.

FS-10615 Everyday Earth Science

EVERYDAY *earth science*

A Far-Out Puzzle

Read each statement below about the puzzle. Using this information, select the correct letter and write it in the numbered space at the bottom of the page. Each clue may refer to more than one letter, so you will have to decide which letter to pick that will spell out a word used in astronomy. Once the word is written, use reference materials to write a short paragraph about the word on the back of this page.

1. The first letter must be in the circle and triangle but not in the square.

2. The second letter must be in the circle and square but not in the triangle.

3. The third letter must be in the circle only.

4. The fourth letter must be in the triangle and square but not in the circle.

5. The fifth letter must be in the square only.

6. The sixth letter must be in the triangle only.

7. The seventh letter must be in the square, below the triangle.

8. The eighth letter must be in the circle, triangle, and square.

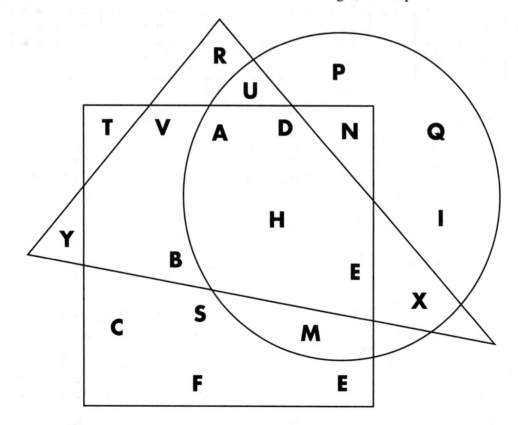

The word is ____ ____ ____ ____ ____ ____ ____ ____
                 1   2   3   4   5   6   7   8

everyday earth science

OUR SUN

The sun is a huge, glowing ball of gases. It is located at the center of the solar system. The words below are related to our sun. Circle the words in the grid. They may be written up, down, forward, backward, or diagonally. For more fun, write the relationship of each topic to the sun.

CHROMOSPHERE	HELIUM	PHOTOSPHERE	SPECTRUM
CORONA	HYDROGEN	PROMINENCE	STAR
FLARES	LIGHT	SOL	SUNSPOTS
HEAT	NUCLEAR	SOLAR WIND	ULTRAVIOLET

```
S  P  R  O  M  I  N  E  N  C  E  B  E  G  C
A  H  C  J  D  H  F  R  S  O  K  I  T  O  H
L  O  S  U  M  X  V  N  T  Y  P  E  R  Q  R
S  T  O  P  S  N  U  S  A  Z  L  O  A  W  O
B  O  N  U  C  L  E  A  R  O  N  C  E  G  M
T  S  D  E  F  H  P  J  I  A  I  M  S  K  O
L  P  O  R  G  Q  S  V  M  S  P  U  O  T  S
U  H  X  A  C  O  A  W  B  Y  V  R  L  N  P
D  E  H  F  L  R  R  I  K  E  G  T  A  E  H
J  R  L  Z  T  O  S  D  Q  S  N  C  R  M  E
L  E  T  L  R  U  E  Z  Y  X  P  E  W  V  R
I  W  U  A  M  U  R  L  E  H  B  P  I  C  E
G  E  D  Y  G  I  A  K  M  J  F  S  N  H  L
H  O  Q  S  U  A  L  E  V  P  G  B  D  W  F
T  Z  C  H  J  T  F  I  D  M  U  I  L  E  H
```

FS-10615 Everyday Earth Science

EVERYDAY EARTH SCIENCE

CONSTELLATIONS

Constellations are the outline of patterns of objects traced among groups of stars in the night sky. The word *constellation* comes from Latin words meaning *together* and *stars*. Even though the stars in a constellation are thought of as belonging to the same group, the stars actually may be extremely far apart.

Pretend that the drawing below represents part of the night sky. Use groups of stars to sketch imaginary constellations. You may use only straight lines to connect the stars, and no lines may cross. Give each constellation a name. The larger dots represent very bright stars. You may also name the stars within a constellation.

● ● ● ● ● ● ● ● ●
E V E R Y D A Y

CONSTELLATIONS
OF THE ZODIAC

Astronomers have divided the sky into 88 constellations. The letters in the blocks below will spell out the names of 12 constellations found in the sky. The beginning letter of each constellation is in the star. Draw straight lines between the letters to find the name of each constellation. No lines will cross. Write the name of each constellation at the bottom of the page.

1.	2.	3.
T T R I ☆S A U S I G A	U A Q R ☆A U I S	E ☆G M I N I
4.	5.	6.
☆C I C A O P R R U N S	O R C O ☆S I P	E C ☆P S I S
7.	8.	9.
S A ☆T U U R	☆C E C N R A	O I ☆V G R
10.	11.	12.
☆L I B A R	☆A R E S I	☆L E O

1. _____   7. _____
2. _____   8. _____
3. _____   9. _____
4. _____   10. _____
5. _____   11. _____
6. _____   12. _____

EVERYDAY EARTH SCIENCE

NORTHERN CONSTELLATIONS

Match the names of the northern constellations with their meanings.

1. _____ Aquarius
2. _____ Aquila
3. _____ Aries
4. _____ Auriga
5. _____ Boötes
6. _____ Cancer
7. _____ Capricornus
8. _____ Coma Berenices
9. _____ Cygnus
10. _____ Draco
11. _____ Gemini
12. _____ Leo
13. _____ Leo Minor
14. _____ Libra
15. _____ Pisces
16. _____ Sagittarius
17. _____ Scorpio
18. _____ Serpens
19. _____ Taurus
20. _____ Ursa Major
21. _____ Ursa Minor
22. _____ Virgo

A. Archer
B. Great Bear
C. Eagle
D. Balance
E. Ploughman
F. Virgin
G. Berenice's Hair
H. Dragon
I. Ram
J. Charioteer
K. Lesser Lion
L. Bull
M. Swan
N. Scorpion
O. Serpent
P. Crab
Q. Little Bear
R. Lion
S. Fishes
T. Twins
U. Goat
V. Water Carrier

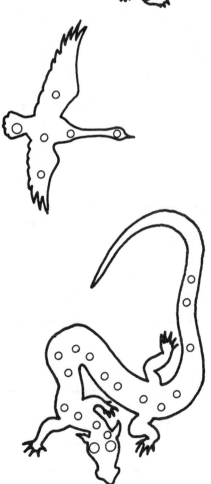

THE BRIGHTEST STARS

The 15 brightest stars of the Northern Hemisphere, as viewed from Earth, are located in some of the best-known constellations. Locate a star map in a reference book or journal and match each star below with its constellation. Some stars appear in the same constellations. The stars are listed in order of brightness.

STAR	CONSTELLATION
1. _____ Sirius	**A.** Taurus
2. _____ Arcturus	**B.** Canis Major
3. _____ Vega	**C.** Orion
4. _____ Capella	**D.** Gemini
5. _____ Rigel	**E.** Boötes
6. _____ Procyon	**F.** Pisces Austrinus
7. _____ Altair	**G.** Canis Minor
8. _____ Aldebaran	**H.** Cygnus
9. _____ Betelgeuse	**I.** Aquila
10. _____ Antares	**J.** Lyra
11. _____ Pollux	**K.** Leo
12. _____ Fomalhaut	**L.** Scorpius
13. _____ Deneb	**M.** Auriga
14. _____ Regulus	
15. _____ Adhara	

Name_____ Date _____

A Giant Red Star

Have you ever seen a bright red star? One really does exist. To find out its name, identify the brightest star in each of the constellations below. Then print the circled letters in the spaces at the bottom of the page.

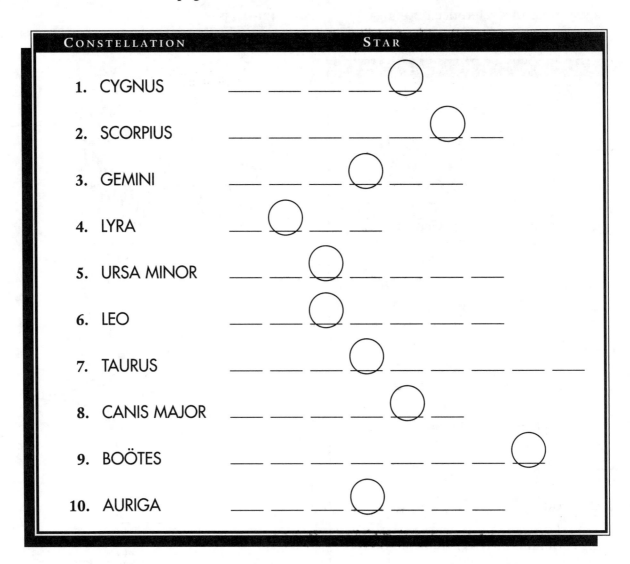

CONSTELLATION	STAR
1. CYGNUS	__ __ __ __ ◯ __
2. SCORPIUS	__ __ __ __ ◯ __
3. GEMINI	__ __ ◯ __ __
4. LYRA	◯ __ __ __
5. URSA MINOR	__ ◯ __ __ __ __
6. LEO	__ __ ◯ __ __ __
7. TAURUS	__ __ ◯ __ __ __ __ __
8. CANIS MAJOR	__ __ __ __ ◯ __
9. BOÖTES	__ __ __ __ __ __ __ ◯
10. AURIGA	__ __ ◯ __ __

11. Name of giant red star: __ __ __ __ __ __ __ __ __ __

12. How do you think the name of this star is pronounced?

_____

CONSTELLATIONS OF SPRING

Circle the names of some of the constellations that can best be seen in the spring sky in the Northern Hemisphere. The words may be written forward, backward, up, down, or diagonally.

AURIGA	CORVUS	HERCULES	LYNX	PUPPIS
BOÖTES	CRATER	HYDRA	ORION	TAURUS
CANCER	GEMINI	LEO	PERSEUS	VIRGO

```
C  A  G  E  S  H  J  T  V  O  E  L  S
R  E  C  N  A  C  T  O  P  U  W  Z  U
A  B  O  F  C  A  T  A  U  R  U  S  E
T  O  R  I  O  N  I  N  P  Y  K  E  S
E  C  V  Q  P  M  O  L  P  L  X  L  R
R  B  U  D  X  B  N  P  I  J  M  U  E
D  E  S  R  O  D  A  B  S  Z  K  C  P
R  T  L  O  W  A  Q  U  H  T  I  R  O
C  S  T  E  A  U  R  I  G  A  Y  E  G
F  E  G  P  O  Q  M  D  S  I  L  H  R
S  H  N  R  U  F  X  N  Y  L  V  K  I
V  G  E  M  I  N  I  G  W  H  J  X  V
```

EVERYDAY EARTH SCIENCE

EVERYDAY EARTH SCIENCE

CONSTELLATIONS OF SUMMER

Circle the names of some of the constellations that can best be seen in the summer sky in the Northern Hemisphere. The words may be written forward, backward, up, down, or diagonally.

AQUILA

BOÖTES

CANES VENATICI

CASSIOPEIA

CEPHEUS

DRACO

HERCULES

LEO

LUPUS

LYRA

OPHIUCHUS

SAGITTARIUS

SCORPIUS

SERPENS

VIRGO

S	S	B	L	U	P	U	S	C	E	S	D	C
A	S	U	S	E	R	P	E	N	S	E	N	A
G	F	C	I	R	A	G	J	M	E	L	Q	S
I	I	P	O	B	A	R	H	S	V	U	W	S
T	O	X	O	R	Q	T	Y	K	I	C	L	I
T	S	D	G	Z	P	A	T	L	C	R	Y	O
A	U	E	R	G	J	I	A	I	O	E	L	P
R	E	H	I	A	T	L	U	T	G	H	U	E
I	H	M	V	S	C	D	I	S	P	A	F	I
U	P	O	Q	B	O	O	T	E	S	K	S	A
S	E	R	N	S	U	H	C	U	I	H	P	O
I	C	I	T	A	N	E	V	S	E	N	A	C

CONSTELLATIONS OF AUTUMN

Circle the names of some of the constellations that can best be seen in the autumn sky in the Northern Hemisphere. The words may be written forward, backward, up, down, or diagonally.

ANDROMEDA	ARIES	DELPHINUS	GRUS	PISCES
AQUARIUS	CAPRICORNUS	DRACO	HERCULES	SCULPTOR
AQUILA	CYGNUS	EQUULEUS	PEGASUS	SERPENS

```
A  E  I  B  R  O  T  P  L  U  C  S  A
C  Q  J  E  H  S  E  C  S  I  P  K  N
F  U  U  I  S  U  N  I  H  P  L  E  D
J  U  L  A  I  K  O  M  C  L  A  N  R
M  L  O  D  R  N  P  G  Y  P  Q  R  O
R  E  S  Q  S  I  U  T  G  V  U  O  M
S  U  R  G  V  Y  U  Z  N  T  W  C  E
X  S  W  P  E  G  A  S  U  S  Y  A  D
A  L  I  U  Q  A  E  B  S  E  I  R  A
S  N  E  P  R  E  S  C  F  X  F  D  I
H  E  R  C  U  L  E  S  D  G  A  G  H
E  C  S  U  N  R  O  C  I  R  P  A  C
```

FS-10615 Everyday Earth Science

EVERYDAY EARTH SCIENCE

CONSTELLATIONS OF WINTER

Circle the names of some of the constellations that can best be seen in the winter sky in the Northern Hemisphere. The words may be written forward, backward, up, down, or diagonally.

ANDROMEDA	CANIS MAJOR	COLUMBA	GEMINI	PERSEUS
AURIGA	CANIS MINOR	ERIDANUS	LEPUS	PISCES
CANCER	CETUS	FORNAX	ORION	TAURUS

```
C O L U M B A B E I F A S
A R E C J S A A O T X B U
N I P Q S G U D R A D P T
I O U E I Z D E C B Y X E
S N S R O J A M S I N A C
M T U I V G I O G R H N G
I A K D F E K R C A E R U
N U N A O M H D W R N P Y
O R J N L I L N U M E G V
R U P U E N X A N R O F Z
T S M S R I D Q W F S G X
P I S C E S U C A N C E R
```

ANIMAL CONSTELLATIONS

Many of the 88 constellations in the sky are named for animals. Match each animal with its constellation.

1. _____ Aquila
2. _____ Aries
3. _____ Camelopardalis
4. _____ Cancer
5. _____ Canis Major
6. _____ Capricornus
7. _____ Cetus
8. _____ Columba
9. _____ Corvus
10. _____ Cygnus
11. _____ Delphinus
12. _____ Draco
13. _____ Equuleus
14. _____ Hydra
15. _____ Leo
16. _____ Lepus
17. _____ Lupus
18. _____ Pegasus
19. _____ Pisces
20. _____ Scorpius
21. _____ Taurus
22. _____ Ursa Major

A.	LION	L.	HARE
B.	DOLPHIN	M.	WHALE
C.	WOLF	N.	COLT
D.	BULL	O.	DOVE
E.	FISH	P.	SCORPION
F.	BEAR	Q.	SEA SERPENT
G.	CROW	R.	EAGLE
H.	HORSE	S.	RAM
I.	GIRAFFE	T.	DOG
J.	GOAT	U.	CRAB
K.	DRAGON	V.	SWAN

Choose one of the constellations listed. Below, draw what you think it looks like up in the sky. Then check an encyclopedia to see if you were close.

EVERYDAY *earth science*

Meanings of Stars' Names

Many stars have Latin, Greek, or Arabic names. Match the meanings of the names with the stars. Then in the box design your own star in the shape of something. Give it an English name and a Latin, Greek, or Arabic version of it.

1. _____ Auriga

2. _____ Algol

3. _____ Antares

4. _____ Arcturus

5. _____ Cepheus

6. _____ Deneb

7. _____ Fomalhaut

8. _____ Gemma

9. _____ Mira

10. _____ Mizar

11. _____ Polaris

12. _____ Procyon

13. _____ Regulus

14. _____ Sirius

15. _____ Spica

A. gem

B. cloak

C. tail

D. ear of grain

E. little king

F. pole star

G. king

H. rival to Mars

I. the demon

J. dog star

K. bear keeper

L. little dog

M. charioteer

N. fish's mouth

O. wonderful

YOUR OWN STAR

Name: _____

_____

_____

PHASES OF THE MOON

The moon seems to change its shape every day as it goes through its phases. Use the code symbols below to identify the phases the moon goes through each month.

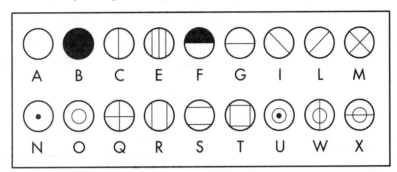

1. ___ ___ ___ ___   ___ ___ ___ ___ ___ ___ ___

2. ___ ___ ___ ___   ___ ___ ___ ___

3. ___ ___ ___ ___ ___ ___   ___ ___ ___ ___ ___ ___ ___

4. ___ ___ ___ ___ ___ ___   ___ ___ ___ ___ ___ ___ ___ ___

5. ___ ___ ___   ___ ___ ___ ___

6. ___ ___ ___ ___ ___ ___   ___ ___ ___ ___ ___ ___

7. ___ ___ ___ ___ ___   ___ ___ ___ ___ ___ ___ ___ ___

8. ___ ___ ___ ___ ___ ___   ___ ___ ___ ___ ___ ___ ___ ___

FS-10615 *Everyday Earth Science*

EVERYDAY EARTH SCIENCE

Name_____ Date _____

JULY 20, 1969

On July 20, 1969, the first person stepped onto the surface of the moon. He said, " . . . one small step for a man, one giant leap for mankind." Identify this person by solving the clues and writing the words in the spaces. The circled letters will then spell out the person's name.

1. Latin word for *moon*   ___ ___ Ⓞ ___

2. Our moon is a _____ of Earth.   ___ ___ ___ Ⓞ ___ ___ ___ ___

3. During a lunar _____, the moon becomes
   dark when it passes through the shadow of Earth.   ___ ___ ___ Ⓞ ___ ___ ___

4. The space mission to the moon was
   called _____.   ___ ___ ___ Ⓞ ___ ___

5. Letters that stand for the organization that conducts research
   into problems of flight within and beyond Earth's atmosphere   ___ Ⓞ ___ ___

6. American space scientists and
   explorers are called _____.   ___ ___ ___ Ⓞ ___ ___ ___ ___ ___

7. The moon's average distance from
   Earth is about 240,000 _____.   Ⓞ ___ ___ Ⓞ

8. The moon's surface is pitted
   with _____.   ___ ___ ___ Ⓞ ___

9. The circling of the spacecraft around
   the moon is called its _____.   ___ ___ Ⓞ ___ ___ ___

10. During a _____ eclipse, the moon
    comes between the sun and Earth.   ___ ___ Ⓞ ___ ___

11. A _____ rocket carried the first
    astronauts to the moon.   ___ ___ ___ ___ ___ Ⓞ

12. The moon's _____ is about
    one-sixth that of Earth's.   Ⓞ ___ ___ ___ ___ ___ ___

The first person to step upon the moon was

___ ___ ___ ___ ___ ___   ___ ___ ___ ___ ___ ___ ___

THE TWO MOONS OF MARS

Mars is a very fascinating planet. It has two moons, or satellites. Complete each sentence about this planet below by writing the missing word or words in the spaces. The circled letters will then spell out the names of the two moons of Mars. Write the names in the spaces at the bottom of the page.

1. Mars is called "the _____ planet."   __ __ ○

2. Dark areas on the surface of Mars are called _____.   __ ○ __ __

                                                              ○

3. Blue and white clouds on Mars are made of _____.   ○ __ __

4. Mars gets as close as about
   35 _____ miles from Earth.   ○ __ __ __ __ __

5. Mars is the _____ planet from the sun.   __ ○ __ __ __

6. For a long time, people thought Mars
   had long channels or _____.   __ __ __ __ __ ○

7. Water _____ has been detected in the
   atmosphere of Mars.   __ __ ○ __ __

8. The space _____ *Mariner VI* passed
   near the surface of Mars and photographed
   the planet's features.   __ __ ○ __ __

9. Ice caps on Mars form at the _____.   __ ○ __ __ __

10. The most common gas on Mars
    is _____ dioxide.   __ __ __ ○ __

11. Deep gorges and _____ have been
    discovered on Mars' surface.   __ __ __ ○ __

12. Much of the surface of Mars is
    pitted with _____.   __ __ __ __ __ ○

The two moons of Mars are

____ ____ ____ ____ ____ ____   and   ____ ____ ____ ____ ____ ____ .

Name _____   Date _____

THE GIANT PLANET JUPITER

Jupiter is the fifth closest planet to the sun. It has 16 satellites, or moons. Identify the name of one of the moons of Jupiter by completing the sentences below. Write the missing words in the spaces to the right. The circled letters will spell out the name of the moon. This moon is actually larger than the planet Mercury.

1. Jupiter is the _____ of all
the planets in our solar system.   ___ ___ ___ ◯ ___ ___ ___

2. Jupiter is crossed with many dark
and light _____.   ___ ◯ ___ ___ ___

3. A day on Jupiter is not
quite _____ hours.   ___ ___ ◯

4. One of the most abundant gases on
Jupiter is _____.   ___ ◯ ___ ___ ___ ___ ___

5. Jupiter gets as close as about
460 _____ miles from the sun.   ◯ ___ ___ ___ ___ ___ ___

6. The diameter of Jupiter is about
_____ times that of Earth.   ___ ___ ◯ ___ ___

7. One of the most interesting features
of Jupiter is its "great _____ spot."   ___ ___ ◯

8. The astronomer _____ discovered
four of the moons of Jupiter.   ___ ___ ___ ___ ___ ◯ ___

A large moon of Jupiter is called   ___ ___ ___ ___ ___ ___ ___ .

ANOTHER NAME FOR VENUS

Venus is a very interesting planet. It even has a nickname. Identify a nickname for Venus by completing the sentences below. Write the missing words in the spaces. The circled letters will spell out this common name. Write the name in the spaces at the bottom of the page.

1. The surface pressure of Venus is 90–100 _____ that of Earth.

    ___ ___ ◯ ___ ___

2. Venus is the _____ planet from the sun.

    ___ ___ ◯ ___ ___ ___

3. The atmosphere of Venus is mostly clouds of _____ dioxide.

    ___ ___ ◯ ___ ___ ___

4. The atmosphere of Venus is so _____ that the surface is very difficult to see.

    ___ ___ ◯ ___ ___ ___

5. Venus is sometimes called "Earth's _____."

    ___ ___ ◯ ___

6. The number of satellites Venus has is _____.

    ___ ___ ◯ ___

7. The temperature on Venus may reach 850 _____ Fahrenheit.

    ___ ◯ ___ ___ ___

8. A day on Venus is 243 of Earth's _____.

    ___ ___ ___ ◯

9. Venus rotates on its axis opposite from all of the other _____.

    ___ ___ ___ ◯ ___

10. There may be some water _____ on Venus.

    ◯ ___ ___ ___

11. Because the light reflected from Venus is so _____, people often think it is a star.

    ___ ◯ ___ ___ ___ ___

A common name for Venus is

___ ___ ___ ___ ___ ___   ___ ___ ___ .

EVERYDAY EARTH SCIENCE

EVERYDAY *earth science*

Our Solar System

There is so much to know about our solar system. To learn a little about it, solve the crossword puzzle below. Use the terms in the box and the clues under the box to help you.

ASTEROIDS	EMIT	LOWELL	MOON	SATURN	TRITON
CERES	HALLEY'S	MARS	NEPTUNE	SPOT	URANUS
COMET	IO	MERCURY	PLUTO	SUN	VENUS
EARTH	JUPITER	METEORS	RINGS	TAIL	WAY

ACROSS

1. rocky particles which orbit the sun mainly between Mars and Jupiter

4. another name for a satellite of a planet

8. the center of our solar system

9. the seventh planet from the sun

11. Planets do not _____ light energy of their own.

13. one of the moons of Neptune

14. sometimes called "the red planet"

15. consists of a head and tail

18. a comet that appears every 76 years

20. known as Earth's twin

21. Saturn is probably best known for its _____.

22. Our solar system is in the galaxy called "The Milky _____."

23. usually the most distant planet from the sun

DOWN

2. the sixth planet from the sun

3. one of the moons of Jupiter

5. the planet discovered in 1846

6. planet covered with dark and light bands

7. planet whose atmosphere is mostly oxygen and nitrogen

10. The Great Red _____ is a prominent feature of Jupiter.

12. often called "falling stars" or "shooting stars"; results when a meteoroid enters Earth's atmosphere from space

15. the largest asteroid

16. planet closest to the sun

17. The _____ of a comet may be over 100 million miles long.

19. American astronomer who began the search for Pluto in 1905

MEMBERS OF OUR SOLAR SYSTEM

Below are descriptions of some members of our solar system. Write the name of the member that matches each description. Then find the answer hidden in the letter blocks. Circle the letters and connect them with straight lines. No lines will cross. Note: The letters can be in circle order or randomly spaced. They will not be found in a straight row or column.

1. a body of ice, gases, and dust which orbits the sun and often has a long tail

_____

A N T U C B L E
I Z C D O N P D
L A U T F M U W
P C X G J P E T

2. the largest planet; has a Great Red Spot

_____

Z J I O P T E C
U B T L M A C U
P I N E B M R S
C A R T V O N E

3. the only planet with plants and animals

_____

C N U I L T A N
B A C M H L R D
E G T X Y F M A
H I U P R M J E

4. a large planet with rings

_____

S U V M E T A I
B A L U I C P O
T M T D R Z M A
W E X A M N E O

5. the belt of thousands of bodies between Jupiter and Mars

_____

C M T L A D U T
I E A S M L C X
T R C D U E L M
M O I U S T A P

6. the closest planet to the sun

_____

X A I P Y L O C
Z T N R L M A R
D F U P A E T J
N Q L C R P A R

7. the sister planet of Earth that is our "evening star"

_____

C M A T U W B Z
A C I S L N R E
N R B D E M W I
A F I S H V A T

8. the last planet to be discovered in 1930

_____

A N T C L R S W
P I N M B Z T U
D L O I Q C B I
M U T C M U E H

9. one of the planets that rotates east to west on its axis

_____

L D R S M I B C
I M U T A E H N
W B N A B O T D
T C U R L F H O

AN ANCIENT ASTRONOMICAL STRUCTURE

One of the best known structures in the world was erected thousands of years ago in southern England. Some people suggest that it enabled the people to study the movements of the sun and moon and to predict eclipses.

Find the name of this structure by using a metric ruler to solve the puzzle below. Each line is represented by a measurement of length in centimeters. Write the answer in the spaces at the bottom of the page.

A 1 cm	**E** 4 cm	**L** 7 cm	**O** 10 cm	**T** 13 cm
C 2 cm	**G** 5 cm	**M** 8 cm	**R** 11 cm	**U** 14 cm
D 3 cm	**H** 6 cm	**N** 9 cm	**S** 12 cm	**W** 15 cm

1. ————————————————

2. ————————————————

3. ———————————————

4. ——————————

5. ————

6. ————

7. ————

8. ——————

9. —————

10. ————

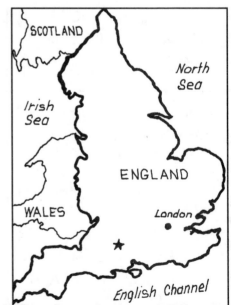

Answer: __ __ __ __ __ __ __ __ __ __

1 2 3 4 5 6 7 8 9 10

MYSTERIOUS STONEHENGE

Stonehenge, an ancient monument built thousands of years ago, in England, has long proved to be a mystery. People have wondered how it was constructed, who built it, and why it was built. Read about Stonehenge in an astronomy or reference book.

```
                              S
                          T   C   B
                  O   L   M   I   F
          N   E   A   C   D   G   S
      E   H   J   M   O   S   E   L   T
  H   T   W   R   Z   X   Q   K   G   B   O
  E   G   N   E   H   E   N   O   T   S   N   P   N
  N   U   A   D   G   E   J   H   E   T   T   I   E   Y   E
  G   K   N   L   F   V   I   N   C   O   G   O   M   O   H   P   H
E   B   Q   X   A   S   B   Z   R   N   I   D   N   W   T   V   E   U   E
S   T   O   N   E   H   E   N   G   E   B   H   J   E   E   G   B   I   N   A   N
T   P   C   K   O   N   C   Y   L   H   U   A   W   M   H   Z   H   E   C   E   O   F   G
R   Q   V   X   F   T   R   G   S   E   I   K   M   H   O   E   J   Q   L   T   R   N   T   S   E
B   U   W   A   Y   D   E   N   V   B   H   F   C   X   N   G   D   Y   Z   P   X   S   S
E   I   K   M   O   G   Q   S   C   J   L   U   R   G   N   V   P   A   W   T   R
D   B   C   E   D   H   G   Y   E   E   I   F   E   X   M   J   L   Y   E
G   P   S   O   N   R   X   Q   G   T   V   Z   U   A   K   W   G
N   D   E   C   S   T   O   N   E   H   E   N   G   E   N
K   J   L   N   I   H   E   G   K   M   F   B   E
E   Q   P   W   R   H   E   G   F   X   H
U   S   O   D   E   T   V   L   E
A   H   Z   N   B   Y   N
I   C   O   K   O
J   T   T
S
```

How many times does the word STONEHENGE appear in the puzzle? The word may be written up, down, forward, backward, or diagonally. Circle the word each time you locate it.

Write five facts about Stonehenge below. Add them to a group "Fact Mural" depicting drawings of and facts about Stonehenge.

1. _____

2. _____

3. _____

4. _____

5. _____

FS-10615 Everyday Earth Science

EVERYDAY EARTH SCIENCE

EARLY DEVELOPMENTS IN ASTRONOMY

Much of the information that is known about astronomy is due to the work of early astronomers, mathematicians, and scientists. Learn about some of these people by matching their names to their appropriate clues below.

A.	ARISTARCHUS	**E.**	KEPLER
B.	ARISTOTLE	**F.**	NEWTON
C.	COPERNICUS	**G.**	PTOLEMY
D.	GALILEO	**H.**	TYCHO BRAHE

1. _____ He proposed that Earth was the center of the solar system and universe.

2. _____ He was the Italian scientist who was the first person to use a telescope for astronomical observations.

3. _____ This Greek astronomer who lived in the 200's B.C. was the first to state that Earth revolves around the sun.

4. _____ A Greek philosopher, he proposed the theory that Earth was the center of the universe. This theory was accepted for about 1800 years.

5. _____ He was a Danish astronomer noted for his recordings of the positions of planets in the sky.

6. _____ He developed the three laws of planetary motion.

7. _____ He was the Polish astronomer who developed the idea of a heliocentric, or sun-centered, system in the 16th century.

8. _____ He formulated the laws of gravitational attraction.

everyday earth science

FAMOUS SPACE SCIENTISTS

Below is a list of scientists who made some wonderful contributions to the space program. Find out what the contributions are by using reference books to locate the information.

SCIENTIST	CONTRIBUTIONS
Edwin Aldrin	
Neil Armstrong	
Yuri Gagarin	
John Glenn	
Robert Goddard	
Fritz Opel	
Sally Ride	
Alan Shepard	
Wernher von Braun	

Name five other famous space scientists. List their contributions to the space programs.

_____

_____

_____

_____

_____

ANSWER KEY

Page 1

Mineral	Hardness	Specific Gravity	Streak Color	Luster
siderite	3.5–4	3.85	white	pearly
gypsum	2	2.32	white	vitreous
kaolinite	2-2.5	2.6	white	dull
halite	2.5	2.16	white	glassy
fluorite	4	3-3.3	white	glassy
calcite	3	2.7	white	waxy
barite	3-3.5	4.3–4.6	white	vitreous
pyrite	6–6.5	5.02	green-black	metallic
galena	2.5-2.7	7.4–7.6	lead-gray	metallic
magnetite	6	5.2	black	metallic
topaz	8	3.4-3.6	colorless	glassy

1. gypsum, kaolinite, halite;
2. gypsum, kaolinite, halite, calcite, galena; **3.** pyrite, magnetite, topaz; **4.** topaz

Page 2

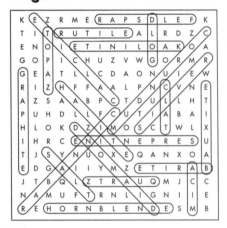

Page 3

1. iron, copper, diamond, gold, graphite, platinum, silver, sulfur;
2. bornite, galena, pyrite;
3. cuprite, hematite, rutile;
4. bauxite, manganite; **5.** fluorite, halite; **6.** calcite, dolomite, malachite; **7.** beryl, feldspar, garnet, mica, olivine, quartz;
8. niter, soda niter; **9.** borax;
10. wolframite; **11.** barite, gypsum; **12.** apatite, turquoise

Page 4

1. talc; **2.** gypsum; **3.** calcite;
4. fluorite; **5.** apatite; **6.** feldspar;
7. quartz; **8.** topaz; **9.** corundum;
10. diamond

Page 5

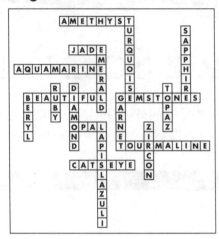

Page 6

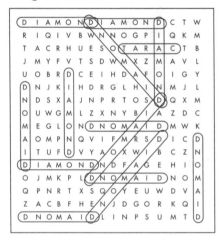

Page 7

Answers will vary.

Page 8

1. aluminum; **2.** iron; **3.** lead;
4. mercury; **5.** copper;
6. chromium; **7.** tin

Page 9

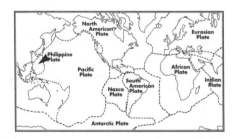

Page 10

1	15	14	4
12	6	7	9
8	10	11	5
13	3	2	16

Page 11

1. tsunami; **2.** seismic;
3. epicenter; **4.** San Andreas;
5. seismograph; **6.** focus;
7. fault; **8.** shock waves; **9.** Ring of Fire; **10.** primary; seismology

Page 12

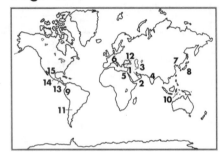

Page 14

3. a. 2730 km; b. 1575 km;
c. 1470 km; d. 630 km; e. 630 km;
f. 1260 km; Missouri; Memphis, St. Louis, Little Rock

Page 16

Alfred Wegener suggested that all continents once were a supercontinent called Pangaea.

Page 17

1. crust; **2.** mantle; **3.** core;
4. caves; **5.** delta; **6.** valley;
7. mountains; **8.** volcano;
9. canyon

Page 18

1. B; **2.** G; **3.** J; **4.** C; **5.** L; **6.** K;
7. N; **8.** I; **9.** M; **10.** E; **11.** F;
12. D; **13.** H; **14.** A

Page 19

1. obsidian; 2. granite; 3. olivine;
4. feldspar; 5. diorite; 6. quartz;
7. basalt; 8. igneous; 9. buildings,
statues, cutting tools, decorative
objects, sand, glass, electronic
equipment; 10. western U.S.,
Mexico, Hawaii, Italy, South
Africa, Canada, North Carolina,
Georgia

Page 20

1. E; 2. H; 3. I; 4. K; 5. M; 6. O;
7. B; 8. N; 9. C; 10. J; 11. G;
12. A; 13. D; 14. L; 15. F

Page 21

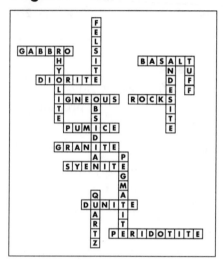

Page 22

Page 23

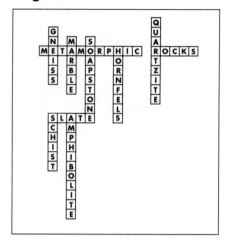

Page 24

Page 25

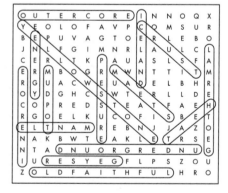

Page 26

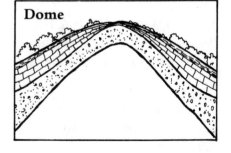

Dome

Fold

Fault-block

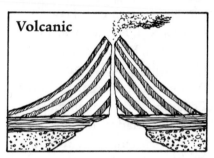

Volcanic

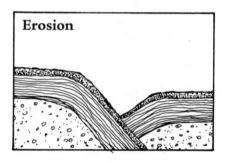

Erosion

Page 27

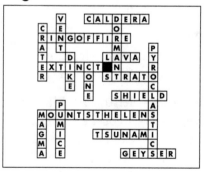

FS-10615 Everyday Earth Science

Page 28
Surtsey, Iceland; Vulcano, Italy; Thera, Greece; Krakatau, Indonesia; Stromboli, Italy; El Chichón, Mexico; Kiska, Alaska; Kilauea, Hawaii; Vesuvius, Italy; Pelee, Martinique; Askja, Iceland; Arenal, Costa Rica; Katmai, Alaska; Beerenberg, Norway; Mayon, Philippines; Pacaya, Guatemala; Poas, Costa Rica; Paricutín, Mexico; Marapi, Sumatra; Mount Baker, Washington; Mount Etna, Italy; Slamet, Java; Fuego, Guatemala; Mount Fuji, Japan; Colima, Mexico; Mauna Loa, Hawaii; Mauna Kea, Hawaii; Galeras, Colombia; Mount Shasta, California; Mount Rainier, Washington; Guagua Pichincha, Ecuador; Tolima, Colombia; Sangay, Ecuador; Ruiz, Colombia; Tupungatito, Chile; El Misti, Peru; Mount Kilimanjaro, Tanzania; Guallatiri, Chile

Page 29
Answers will vary.

Page 30
1. period; **2.** bacteria; **3.** years; **4.** Grand Canyon; **5.** mountains; **6.** million; **7.** billion; **8.** meteorites; **9.** fossils; **10.** Paleozoic; **11.** longest; Precambrian

Page 31
Eras—Cenozoic, Mesozoic, Paleozoic, Precambrian

Periods—Pleistocene, Pliocene, Miocene, Oligocene, Eocene, Paleocene, Cretaceous, Jurassic, Triassic, Permian, Pennsylvanian, Mississippian, Devonian, Silurian, Ordovician, Cambrian

Page 32
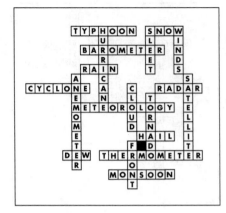

Page 33
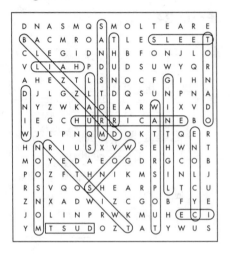

Page 34

1	15	14	4
12	6	7	9
8	10	11	5
13	3	2	16

34; 4 squares in each corner, 4 squares in center

Page 35
1. b; **2.** a; **3.** d; **4.** d; **5.** c; **6.** c; Weather satellites are valuable in predicting hurricane movements.

Page 36
1. radiosonde; **2.** doppler; **3.** ceilometer; **4.** rain gauge; **5.** radar; **6.** anemometer; **7.** vane; **8.** barometer; **9.** thermometer; **10.** thermograph; **11.** hygrometer; solar energy

Page 37
Answers will vary.

Page 38
Relative Humidity: 83%; 12%; 65%; 30%; 69%; 90%; 31%

Page 39
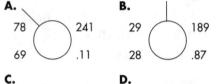
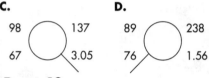

Page 40
doldrums, trade winds, prevailing easterlies, prevailing westerlies, horse latitudes

Page 41
1. cirrus; **2.** cirrostratus; **3.** altocumulus; **4.** stratus; **5.** cumulus; **6.** nimbostratus

Page 42
1. F; **2.** T; **3.** T; **4.** T; **5.** F; **6.** T; **7.** F; **8.** T; **9.** F; **10.** T; **11.** T; **12.** T; **13.** T; **14.** T; **15.** T; **16.** T; **17.** T; **18.** T; **19.** T; **20.** T; **21.** T; **22.** F; **23.** F; **24.** F

Page 43
Answers will vary.

Page 44
Answers will vary.

Page 45
Answers will vary.

Page 46
weather instruments—wind vane, barometer, thermometer, anemometer, radar, hygrometer; types of precipitation—rain, sleet, snow, dew, hail; layers of the atmosphere—troposphere, stratosphere, thermosphere, mesosphere, ionosphere; weather map—cold front, warm front, low pressure, isobar, high pressure; types of clouds—stratus, cirrus, cumulus, cumulonimbus; storm systems—monsoon, tornado, thunderstorm, typhoon, hurricane

Page 47
Answers will vary.

Page 48
1. T; **2.** F; **3.** F; **4.** T; **5.** F; **6.** F; **7.** T; **8.** F; **9.** T; **10.** T; **11.** T; **12.** T; **13.** T; **14.** F; **15.** F; **16.** F; **17.** F; **18.** F; **19.** T; **20.** F; **21.** T; **22.** T; **23.** F; **24.** T

Page 49
25. T; **26.** F; **27.** F; **28.** T; **29.** F; **30.** F; **31.** T; **32.** F; **33.** T; **34.** F; **35.** T; **36.** T; **37.** T; **38.** F; **39.** T; **40.** T; **41.** T; **42.** T; **43.** T; **44.** F

Page 50
1. Hey from Earth

Page 51
universe

Page 52
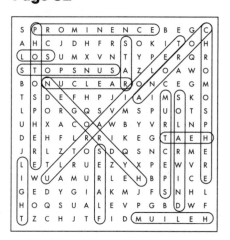

Page 53
Answers will vary.

Page 54
1. Sagittarius; **2.** Aquarius; **3.** Gemini; **4.** Capricornus; **5.** Scorpio; **6.** Pisces; **7.** Taurus; **8.** Cancer; **9.** Virgo; **10.** Libra; **11.** Aries; **12.** Leo

Page 55
1. V; **2.** C; **3.** I; **4.** J; **5.** E; **6.** P; **7.** U; **8.** G; **9.** M; **10.** H; **11.** T; **12.** R; **13.** K; **14.** D; **15.** S; **16.** A; **17.** N; **18.** O; **19.** L; **20.** B; **21.** Q; **22.** F

Page 56
1. B; **2.** E; **3.** J; **4.** M; **5.** C; **6.** G; **7.** I; **8.** A; **9.** C; **10.** L; **11.** D; **12.** F; **13.** H; **14.** K; **15.** B

Page 57
1. Deneb; **2.** Antares; **3.** Castor; **4.** Vega; **5.** Polaris; **6.** Regulus; **7.** Aldebaran; **8.** Sirius; **9.** Arcturus; **10.** Capella; **11.** Betelgeuse; **12.** Beetle Juice

Page 58
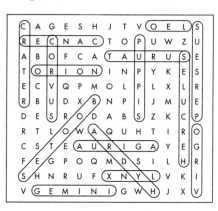

Page 59
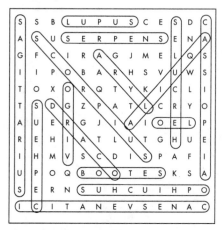

Page 60
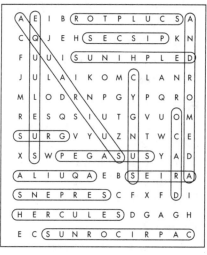

Page 61
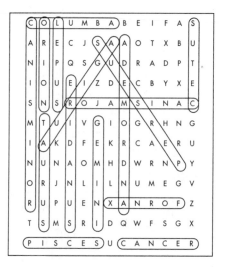

Page 62
1. R; **2.** S; **3.** I; **4.** U; **5.** T; **6.** J; **7.** M; **8.** O; **9.** G; **10.** V; **11.** B; **12.** K; **13.** N; **14.** Q; **15.** A; **16.** L; **17.** C; **18.** H; **19.** E; **20.** P; **21.** D; **22.** F

Page 63
1. M; **2.** I; **3.** H; **4.** K; **5.** G; **6.** C; **7.** N; **8.** A; **9.** O; **10.** B; **11.** F; **12.** L; **13.** E; **14.** J; **15.** D

Page 64
1. last quarter; **2.** full moon; **3.** waxing gibbous; **4.** waning crescent; **5.** new moon; **6.** waning gibbous; **7.** first quarter; **8.** waxing crescent

FS-10615 Everyday Earth Science

Page 65
1. luna; **2.** satellite; **3.** eclipse;
4. *Apollo;* **5.** NASA; **6.** astronauts;
7. miles; **8.** craters; **9.** orbit;
10. solar; **11.** Saturn; **12.** gravity;
Neil Armstrong

Page 66
1. red; **2.** seas; **3.** ice; **4.** million;
5. fourth; **6.** canals; **7.** vapor;
8. vehicle; **9.** poles; **10.** carbon;
11. canyons; **12.** craters;
Deimos, Phobos

Page 67
1. largest; **2.** bands; **3.** ten;
4. hydrogen; **5.** million;
6. eleven; **7.** red; **8.** Galileo;
Ganymede

Page 68
1. times; **2.** second; **3.** carbon;
4. dense; **5.** twin; **6.** none;
7. degrees; **8.** days; **9.** planets;
10. vapor; **11.** bright;
Morning Star

Page 69

Page 70
1. comet; **2.** Jupiter; **3.** Earth;
4. Saturn; **5.** asteroids;
6. Mercury; **7.** Venus; **8.** Pluto;
9. Uranus

Page 71
Stonehenge

Page 72

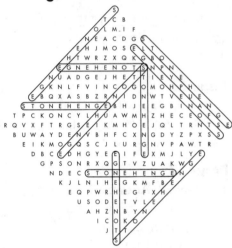

Page 73
1. G; **2.** D; **3.** A; **4.** B; **5.** H; **6.** E;
7. C; **8.** F

Page 74
Aldrin—U.S. astronaut, made
first lunar landing in 1969;
Armstrong—U.S. astronaut,
made first lunar landing in
1969, first person to step on
moon; Gagarin—Russian
cosmonaut, first man to be
launched in a spacecraft in
1961; Glenn—U.S. astronaut,
first American to orbit Earth in
a spacecraft in 1962; Goddard—
American physicist whose
experiments on rockets between
1909–1945 led to present-day
rockets; Opel—inventor who
experimented with dry-fuel
rockets in the 1920–1930s;
Ride—first female U.S.
astronaut; Shepard—first U.S.
astronaut to make a suborbital
flight in a spacecraft in 1961;
von Braun—German rocket
scientist who came to the U.S.;
He directed teams that built
the rockets that sent the first
Americans into space and to the
moon; He also developed the
Redstone rocket.

FS-10615 Everyday Earth Science